The Official
BBC micro:bit®
User Guide

Gareth Halfacree

About the Author

GARETH HALFACREE is a freelance technology journalist and the co-author of the *Raspberry Pi User Guide* alongside project co-founder Eben Upton. Formerly a system administrator working in the education sector, Gareth's passion for open source projects has followed him from one career to another, and he can often be seen reviewing, documenting or even contributing to projects such as GNU/Linux, LibreOffice, Fritzing and Arduino. He is also the creator of the Sleepduino and Burnduino open hardware projects, which extend the capabilities of the Arduino electronics prototyping system. A summary of his current work can be found at `http://freelance.halfacree.co.uk`.

About the Technical Editor

DAVID WHALE is an embedded software engineer whose career of over 30 years has involved him designing and building embedded software for a diverse collection of high tech products. David is a STEM Ambassador and volunteer for the Institution of Engineering and Technology (IET), where he regularly helps schools and teachers introduce and run computing clubs, as well as trains teachers nationally. It was through his association with the IET that David became involved with the micro:bit project, where he has helped to develop a large base of teaching resources, as well as train and support the delivery of the IET Faraday national STEM challenge days using the BBC micro:bit. David now works with the Micro:bit Educational Foundation, where he continues to develop projects and resources with many partner organisations, which includes the Doctor Who team at the BBC. David is the co-author of the successful Wiley title *Adventures in Minecraft,* a book that teaches Python coding to children via their interest in the Minecraft game, and he has been technical editor of a wide range of technology and computing books.

Credits

Project Editor
John Sleeva

Technical Editor
David Whale

Production Editor
Barath Kumar Rajasekaran

Copy Editor
Karen A. Gill

Production Manager
Katie Wisor

Manager of Content Development and Assembly
Mary Beth Wakefield

Marketing Manager
Christie Hilbrich

Professional Technology & Strategy Director
Barry Pruett

Business Manager
Amy Knies

Executive Editor
Jody Lefevere

Associate Acquisitions Editor
Riley Harding

Project Coordinator, Cover
Brent Savage

Proofreader
Debbye Butler

Indexer
Estalita M. Slivoskey

Cover Designer
Wiley

Cover Image
Courtesy of Gareth Halfacree

Contents

Part III
CHAPTER 8
The Wireless BBC micro:bit 163

CHAPTER 9
The BBC micro:bit and the Raspberry Pi 175

CHAPTER 10
Building Circuits 193

CHAPTER 11

CHAPTER 12

Foreword

IN APRIL 2015, I spent a lunchbreak searching my local supermarket for the cheapest thing I could cannibalise to finish a demo for the BBC. What they saw that afternoon was the battery clip from a £1 alarm clock, glued onto an early BBC micro:bit prototype. I was building a demo as one of the 31 partner organisations that the BBC had recruited to help them deliver a bold and audacious project: to give a coding device to every year 7 in the UK, for free. However, the BBC micro:bit wasn't designed to be just another programmable 'development board', but a seamless plug-and-play tool that puts creativity, learning, and ease-of-use for teachers and young people first.

As an engineer on the project, the most compelling thing about working with the BBC micro:bit is seeing the exciting (and sometimes ridiculous) things these new audiences choose to do with this technology: build a team game based on a teleporting duck, measure a rocket car's acceleration, tell the interactive story of pizza, build a portable heart-rate-monitor, or invent the fantastic 'rain detecting hat' you can find in this book. This guide brilliantly captures the exhilaration, simplicity, and creative potential of the BBC micro:bit, and I'm sure it will help many more people become coders and inventors.

Instantly interactive, the BBC micro:bit's sensors and slick design make it accessible and exciting to the widest possible audience, even (perhaps especially) those who didn't realise that coding was for them. It takes no time at all to build your first program, and the simplicity of the tools means that what follows is an upward spiral of success and satisfaction that encourages your imagination to run wild.

Part of the magic of the BBC micro:bit is how effortlessly it brings the virtual and physical together. Within minutes, things you've made with the computer start controlling the real world. It's this physicality and immediacy that create the 'micro:bit moment'. It's amazing to see people of all ages have that empowering experience when they realise the potential of the device in their hands, and want to make something new with it—the moment people start to feel excited about playing with technology instead of nervously holding back from it.

But micro:bit isn't really about learning to code; it's about learning to do things that you care about with technology. Learning to code happens along the way: you learn about loops almost by accident because you're making your BBC micro:bit dance, and the song has a verse and chorus that repeat. This approach helps engage new coders of all ages. Independent research tracked the progress of the BBC micro:bit in the UK during its first year and found that 85% of teachers who used the device agreed that it made Computer Science more enjoyable for their students, and 90% of the children who used it said it helped show them that anyone can code.

The cool design, broad appeal, and technological simplicity of the micro:bit tools come as a direct result of the diverse backgrounds, outlooks, communities, and interests within the micro:bit partnership. This broad, interdisciplinary team has shaped the device, the editors, the learning materials, and the concept of the micro:bit itself into a truly unique ecosystem.

Founded in September 2016, the Micro:bit Educational Foundation has been entrusted with supporting and developing that ecosystem, and taking the micro:bit project around the world. At the time of writing, the device is available in more than 50 countries, with the website available in 12 languages. We have a growing library of lesson plans, projects and ideas, new, more advanced editors developed with our partners (which are explained in this book), and a growing community of enthusiasts, volunteers, great partners, and educators.

So, getting started with the BBC micro:bit isn't just about building your own creations. It's also about joining a worldwide community of people who are using technology creatively to express themselves, solve problems they care about, make life better for others, and help change the way it feels to learn to code.

This book will introduce you to the world of micro:bit, but it's only the start of your journey. Welcome to the micro:bit community. We're looking forward to seeing what you create!

—Jonny Austin, CTO, Micro:bit Educational Foundation

Introduction

EDUCATIONAL COMPUTING HAS a long and storied history, beginning with the adoption of mechanical calculators to aid with mathematics classes through to the early days of microcomputing with initiatives like the BBC Computer Literacy Project in the 1980s. As the cost of computers came down and their capabilities increased, schools around the world rapidly went from a single shared computer to entire rooms filled with computers, integrating them into lessons from languages and history to engineering and art.

Today many homes have a computer of their own, or in some cases more than one. While access to computers has increased, actually operating them brings with it a sense of being disconnected from their inner workings. The BBC Micro, the 1980s microcomputer designed by Acorn Computers and at the heart of the BBC Computer Literacy Project, loaded straight into a text-based programming language known as the Beginner's All-Purpose Symbolic Instruction Code (BASIC) and invited experimentation; today, the majority of computers load into a graphical user interface (GUI) which emphasises the use of pre-written programs over creating your own.

The BBC micro:bit is designed to bring back the days of learning to write your own code on a low-cost, easily-understood platform. Designed to sit at the heart of an international computer literacy programme directly inspired by the BBC's original Computer Literacy Project, the BBC micro:bit is an affordable microcontroller on which you can run your own programs to do everything from spell out your name or play a game to turn lights on and off or communicate via radio.

Schools around the world have begun to adopt the BBC micro:bit in their curricula, but it's by no means exclusively for structured educational use. The device's readily accessible nature means it can be used to teach programming and computing concepts to children of any age, its robustness and small size mean it's extremely portable and well-suited to wearable projects, and its surprising power and flexibility mean that you're unlikely to find yourself restricted by its capabilities even when using it at the heart of more complex projects.

Who This Book Is For

This book is written for anyone interested in working with the BBC micro:bit. You don't need pre-existing knowledge of computers, electronics, or programming to be able to pick up a BBC micro:bit and get started.

All you need to get the most from this book is the ability to read and a willingness to learn. If you've used computers before, you'll find that you are able to skim through some of the early chapters on general concepts. If you've used other microcontroller-based development boards, you can skip still more. If you've written your own computer programs, then you'll find programming for the BBC micro:bit immediately familiar. And if you know your way around an electrical circuit, the later chapters should hold few surprises.

Whether you're an existing user of the BBC micro:bit or a complete newcomer, this book aims to get you started on your journey with a minimum of fuss and maximum enjoyment.

What This Book Covers

The march of technology is constant, and the BBC micro:bit is no exception. This book has been written based on the most recent version of the BBC micro:bit hardware, revision 1.3b, but it is entirely applicable to all versions going back to the first prototype versions given to schools for testing purposes. Equally, it will remain applicable to all future revisions thanks to the sterling work of the Micro:bit Educational Foundation, which has been instrumental in the development of this book.

The software for the BBC micro:bit is, as with the hardware, under constant development. References to the software in this book are accurate at the time of writing, and all screen-shots and related materials have been captured on the very latest software versions. Over time, small changes may be made to the way the software looks, but the way it works will remain the same.

This is especially important for the programs contained within the book. Although the languages available for the BBC micro:bit will expand over time and gain additional features, the existing features will always be available. A program taken from this book today will still be usable with the BBC micro:bit years down the line.

How This Book Is Structured

Part I, 'An Introduction to the BBC micro:bit', offers a guide to the hardware and how it works, step-by-step instructions on unpacking your first BBC micro:bit and exploring its sample program, and connecting the BBC micro:bit to your computer so you can load a program of your own. This section also contains a number of tips on working with the BBC micro:bit, including how to handle it to protect it against possible damage. Even if you've already worked with a BBC micro:bit, reading this section is well advised.

Part II, 'Coding for the BBC micro:bit', gets you started writing your own programs. You learn the languages available for the BBC micro:bit and how they differ, and you learn to write your own programs from a simple message scrolling across the BBC micro:bit's display and reading from its various built-in sensors to writing a simple game.

There's a chapter dedicated to each of the three main programming languages used with the BBC micro:bit: JavaScript Blocks, JavaScript, and Python. Each chapter is designed to be as close to identical to the others as possible, allowing you to quickly see how the process of writing each program differs between languages. You can use the comparison table at the start of the section to pick a favoured language and read only that language's chapter, or you can work through all three chapters in turn to get a real feel for how each operates.

Part III, 'Advanced BBC micro:bit Projects', goes a step further, introducing the radio module with examples on communicating between individual BBC micro:bits and groups of BBC micro:bits without the need for wires. There's also a chapter dedicated to using the BBC micro:bit with the popular Raspberry Pi educational single-board computer, extending the capabilities of both devices.

In addition, you learn how to add external components like switches and LEDs to the BBC micro:bit, building electronic circuits from basic components to further extend its functionality. There's no soldering involved, and the circuits described are designed to be safe and accessible for even the youngest reader; they require only a small number of affordable electronic components.

You will now have a sound understanding of how the BBC micro:bit works, how to program it, and how to use it with other devices. You still won't have reached the end of its capabilities, though, so the final chapter offers information on additional resources, including add-on hardware which can further increase the BBC micro:bit's flexibility, and websites offering more project ideas and formal lesson plans for use in structured education.

Finally, the appendices have full program listings for every program mentioned in the book in all three languages, making it easy to type them in without getting distracted by comments and explanations of what each part of the program is doing. If you'd rather save your fingers, you can download the same program files from the book's website at `www.wiley.com/go/bbcmicrobituserguide`. You'll also find a *pin-out* diagram of the BBC micro:bit with a full list of its capabilities.

What You Need to Use This Book

Technically speaking, you can begin using this book even without a BBC micro:bit of your own; simulators allow you to write programs designed for the BBC micro:bit and see how they run even without loading them onto a physical BBC micro:bit. You'll get the most from the book if you have at least one BBC micro:bit with which to experiment, however, along with a few extras detailed here.

To run the main program samples listed in this book, you need the following:

- A BBC micro:bit
- A fully-wired micro-USB cable
- A computer running Microsoft Windows, Apple macOS, or Linux, with a free USB port
- An up-to-date web browser and working Internet connection

For the radio programs included in Chapter 8, 'The Wireless micro:bit', you also need:

- A total of three BBC micro:bits

To run the programs listed in Chapter 9, 'The BBC micro:bit and the Raspberry Pi', you need:

- A Raspberry Pi Model B+, Raspberry Pi Model 2, Raspberry Pi Model 3, or Raspberry Pi Zero W
- A micro-USB On-The-Go (OTG) adapter cable, if using the Raspberry Pi Zero W

To build the circuits detailed in Chapter 10, 'Building Circuits', you also need the following:

- Wires with crocodile clip or 4mm banana plug connectors
- A button or switch
- An LED
- A current-limiting resistor (see Chapter 10 for an explanation)
- A potentiometer

These parts are readily available via the Internet or in high-street electronics component shops, and they can frequently be supplied by BBC micro:bit resellers alongside the BBC micro:bit itself.

Conventions

To help you get the most from the text and keep track of what's happening, we've used a number of conventions throughout the book.

Technical terms are presented *in italic* when they're first used. The same applies to acronyms and initialisms, which are presented in full when first used and then in their abbreviated form.

Metric measurements are used throughout this book, with imperial measurements provided in brackets where appropriate.

When a line of code would extend past the border of the page, a ↩ symbol is printed. When you see this symbol, continue to type the code without pressing the Enter or Return keys. If you're not sure how a line of code should be entered, visit the website at `www.wiley.com/go/bbcmicrobituserguide` to download plain-text versions of each program; these can then be used for reference or even simply copied and pasted directly into the editors.

Contact Me

Comments, corrections, and questions from readers are heartily welcomed via email at microbit@halfacree.co.uk, while other publications of mine can be found at `freelance.halfacree.co.uk`.

You can also get in touch with me via Twitter at `twitter.com/ghalfacree` and via encrypted message at `keybase.io/ghalfacree`.

Enjoy the book, and happy travels on your BBC micro:bit journey!

—Gareth Halfacree

Part I

An Introduction to the BBC micro:bit

Chapter 1
Meet the BBC micro:bit

In this chapter

- A look at what the BBC micro:bit is and how it came about
- A tour of the BBC micro:bit and an explanation of its major components

THE BBC MICRO:BIT is an incredible device, capable of educating and entertaining in equal measure. It can form the heart of a complex robotic or home automation system, or it can simply show a smiley face when you press a button. It can help you come to grips with programming, learn about how electronics work, and even communicate wirelessly with more BBC micro:bits or other devices such as a smartphone or tablet.

You can write programs for the BBC micro:bit in a variety of *programming languages*, or you can use programs others have written. You can use the BBC micro:bit in the classroom, the club, the playground, or at home. You can play games, solve problems, and invent new devices, all with your BBC micro:bit.

Before all this, though, you'll need to meet the BBC micro:bit.

A Tour of the Board

The BBC micro:bit is what is technically known as a *microcontroller development board*. That is, it's a *printed circuit board* (PCB) which contains a *microcontroller* on which you can run your own programs and connect your own hardware.

The first microcontroller development boards were expensive and complicated to use. In the decades since their first introduction, they have become steadily cheaper and more accessible, until the BBC micro:bit became possible: a minimal-cost, highly-functional board designed to help teach programming, or 'coding', to anyone regardless of experience.

Your programming journey begins simply: learning about the BBC micro:bit itself.

WARNING The BBC micro:bit is designed to be robust, but it's still a complex electronic device. It's designed as a bare circuit board so that you can see what all its components are and what they do, but this does mean you need to take a little more care with handling it than if it were in an enclosure. Always make sure to handle it by its edges to avoid damage through *electrostatic discharge*. For more information on preventing electrostatic discharge damage, see Chapter 2, 'Getting Started with the BBC micro:bit'.

The BBC micro:bit itself is a small printed circuit board with a range of components fitted. It has two sides: the front side of the BBC micro:bit includes the *display* and *buttons*, while the back side has components like the *micro-USB connector* and *radio*. Figure 1-1 shows the front side of the BBC micro:bit, and Figure 1-2 shows the back.

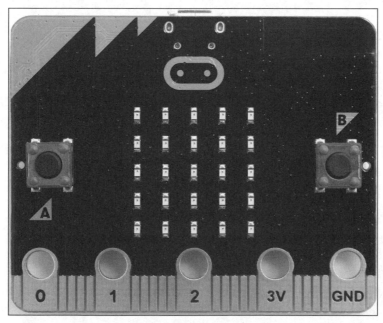

FIGURE 1-1: The front of the BBC micro:bit

FIGURE 1-2: The back of the BBC micro:bit

The BBC micro:bit includes a layer of print known as a *silkscreen layer*, which labels some of the more important components. The buttons on the front side are labelled A and B so you know which is which, while the back side picks out key components like the *processor* and the *accelerometer*. These components, and others, are described in more detail in the next section.

Breaking It Down

Like any complex device, the BBC micro:bit is made up of numerous relatively simple parts. From the more obvious, like the display which dominates the front of the board, to the more subtle, like the radio which allows two or more BBC micro:bits to communicate with each other or connect to a smartphone or tablet, each component works with the others to give the BBC micro:bit its impressive flexibility.

If you're impatient to get started with your BBC micro:bit, you can skip ahead to Chapter 2. Reading the rest of this chapter, though, will give you a good understanding of just what the BBC micro:bit can do, putting you in good stead for not only learning how it works but coming up with practical projects of your own as your skills progress.

Display

The most obvious feature of the BBC micro:bit is its display, which sits in the centre of the board's front side (see Figure 1-3). This is the BBC micro:bit's primary *output device*, a means for a program running on the BBC micro:bit to communicate with the outside world— whether that's reading off the position of one of the BBC micro:bit's sensors or simply displaying a smiling face.

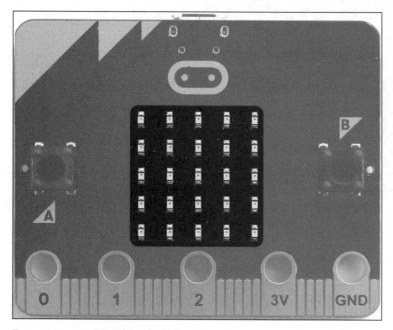

FIGURE 1-3: The BBC micro:bit's display

The BBC micro:bit's display serves the same purpose as the monitor on a desktop computer but is of a considerably lower *resolution*. Where a desktop monitor may be made up of 1,920 columns of 1,080 picture elements or *pixels*, the BBC micro:bit has just five columns of five picture elements for a total of 25 pixels.

Although 25 pixels may not sound like much, it's more than enough to view useful information such as a bar graph or the elements of a simple game or to scroll messages of almost any length. You'll learn more about using the display for these tasks and more later in the book.

From a technical perspective, the BBC micro:bit's display is made up of a 5×5 *light emitting diode (LED) matrix*. Each LED in the matrix makes up one pixel and can display a single colour in varying brightnesses. By altering the brightness and rapidly changing the image being shown, the BBC micro:bit's display can show animation as well as still images.

Buttons

Next to the display, the buttons are the BBC micro:bit's second most obvious feature. The two main buttons, Button A and Button B, are positioned either side of the display at the front of the board and are labelled with their letter to avoid any confusion (see Figure 1-4). These two buttons form the BBC micro:bit's primary *input devices*. Where the display allows information to be output from a program running on the BBC micro:bit, the buttons allow you to send simple inputs into the program to change the image being displayed, for example, or control a character in a game.

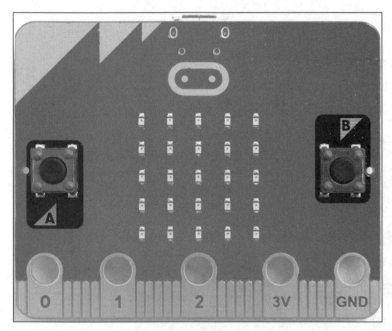

FIGURE 1-4: The BBC micro:bit's Button A and Button B

These buttons are technically known as a *momentary switch*. These are different to the switches you might use to turn on a light, which are known as *latching switches* or *toggle switches*. A momentary switch stays 'on' only as long as you continue to push it down, while a latching switch goes from 'off' to 'on' and stays there until you push it again.

Although the buttons are permanently connected to the BBC micro:bit, they won't do anything unless the program running on the BBC micro:bit is listening for a button input. Depending on what you're using the BBC micro:bit for, you may want to use only one or even neither of the two buttons; alternatively, you may find you need more buttons, in which case the *input-output pins* will help you expand the BBC micro:bit.

In addition to Button A and Button B, there's a third button on the back of the BBC micro:bit: the *Reset button*. Like the Reset button of a desktop computer, the BBC micro:bit's Reset button acts like cutting the power. Whatever the BBC micro:bit is doing at the time will be interrupted, and the BBC micro:bit will restart and begin its stored program again from the start. The Reset button is useful when experimenting with programs that may cause the BBC micro:bit to stop responding, but be careful not to press it accidentally when you're in the middle of something. There are other uses for the Reset button, too, which will be explained in later chapters.

Processor

The *processor* is often called the 'brain' of a computer, and the BBC micro:bit's processor—technically known as a *microcontroller*—is no exception. Found on the upper-left of the rear of the BBC micro:bit and labelled on the board's *silkscreen layer*, the processor is a tiny black square no bigger than your little fingernail called an *integrated circuit* (see Figure 1-5).

FIGURE 1-5: The BBC micro:bit's processor

Despite its small size, this integrated circuit is surprisingly complex. It's here that any program the BBC micro:bit is using is both stored and run. While in a desktop computer, memory, storage, and the central processing unit are all separate, the BBC micro:bit's processor is all-in-one; this is known as a *system-on-chip* (SoC).

The BBC micro:bit's processor uses a special set of instructions, known as an *instruction set architecture*, called the *ARM Architecture*. Named for the company which invented it, ARM processors are designed to offer high performance at the lowest possible power usage. The BBC micro:bit uses this to great effect: It's possible to run the BBC micro:bit for months when using the recommended batteries.

For those interested in the technical side, the BBC micro:bit's processor is a Nordic nRF51822, which contains a single-core ARM Cortex-M0 running at 16Mhz, with 16 KB of random access memory (RAM) and 256 KB of non-volatile memory (NVM) for program storage.

The BBC micro:bit also has a second processor, not labelled on the silkscreen and found at the upper-right of the board. When you connect the BBC micro:bit to your computer using a micro-USB cable, as described in Chapter 3, 'Programming the BBC micro:bit', this second processor takes over and handles communication with your computer, accepting new programs and transferring them to the main processor to run.

Radio

A major feature of the BBC micro:bit is its *radio*, which allows it to communicate with other BBC micro:bits or with other devices, such as a smartphone or tablet. The radio itself is a part of the BBC micro:bit's main processor, forming a segment of the system that makes up the system-on-chip design. As a result you won't find it labelled on the silkscreen as with other components, but instead covered under the 'PROCESSOR' label.

The BBC micro:bit's radio has two main functions. The first function is to communicate with other BBC micro:bits wirelessly, allowing you to group multiple BBC micro:bits without having to string cables between them. The second function is to communicate with other devices, which it does using *Bluetooth Low Energy (BLE)*, a low-power version of the popular Bluetooth wireless standard built into most modern smartphones and tablets.

Unlike the type of radio you might use to listen to music, there's no external antenna for the BBC micro:bit's radio chip. Instead, it uses a cleverly-shaped copper track built into the circuit board itself. You can find this at the upper-left on the rear of the BBC micro:bit, labelled 'BLE ANTENNA' (see Figure 1-6). It doesn't look like much, but if you hold it to the light, you should be able to see a raised line tracing a series of rectangular shapes—providing, that is, the antenna isn't covered by regulatory compliance stickers required of all radio transmitters in selected countries.

FIGURE 1-6: The BBC micro:bit's radio antenna

Accelerometer

One of two built-in sensors, the BBC micro:bit's *accelerometer* is an integrated circuit even smaller than the processor chips (see Figure 1-7). It's so small that if it weren't labelled on the board's rear silkscreen, it'd be easy to overlook it. Despite its small size, this chip is extremely clever: it knows exactly how the BBC micro:bit is positioned in space at any given time.

When you rotate your smartphone from portrait to landscape and vice versa, it's an accelerometer that tells the device what you've done and allows it to automatically rotate the screen. The BBC micro:bit's accelerometer works in the same way: it can track the angle of the device in all three *axes*—sideways, backward and forward, and up and down, or X, Y, and Z—by tracking what is known as *proper acceleration*.

The accelerometer can be seen in action during the BBC micro:bit's demonstration program, which is explored in Chapter 2.

FIGURE 1-7: The BBC micro:bit's accelerometer

Compass

The second of the BBC micro:bit's two built-in sensors, the *compass*, works roughly like the navigational tool of the same name: the compass will detect *magnetic north* and point you in that direction. If you're building a robot, for example, the compass is a reliable method of navigating between points. As with the accelerometer, the compass is a tiny and easily-overlooked integrated circuit on the rear of the device and is labelled on the silkscreen (see Figure 1-8).

Like a traditional needle-based compass, the BBC micro:bit's compass works by sensing *magnetic fields*. As a result, it has a second trick up its sleeve: the ability to detect magnetic fields other than the Earth's natural ones. Using this chip, the BBC micro:bit can not only point the way north but can also indicate the strength of a local field's *magnetic force*—and even detect metal. As it works by detecting these magnetic fields, however, this means that nearby magnets—such as those found in speakers—can affect the sensor's accuracy when used as a compass.

FIGURE 1-8: The BBC micro:bit's compass

Input-Output Pins

The BBC micro:bit has room for expansion in the form of *input-output pins* located on the bottom edge (see Figure 1-9). These aren't literally pins but strips of copper on both sides of the BBC micro:bit's printed circuit board; the term 'pin' is a technical term referring not to the connections themselves but to the 'pin' of the processor to which the connections are linked.

The five largest pins, sometimes referred to as *pads* or *rings* in reference to their shape, are labelled on the front of the board: 0, 1, 2, 3V, and GND. The first three are the BBC micro:bit's major input output pins themselves, while the latter two provide power and a ground connection to complete any circuit you build. Each of these pins have a small hole at the top, which allows you to quickly connect hardware to the BBC micro:bit using *crocodile clip* or *banana plug* connectors. They're also suitable for use with *conductive thread* (see Chapter 12, 'The Wearable BBC micro:bit') and sometimes via small conductive screws, as with add-on boards (see Chapter 11, 'Extending the BBC micro:bit').

As the name suggests, the input-output pins can be used for either sending an input to the BBC micro:bit or taking an output from the BBC micro:bit. You could connect a temperature sensor to Pin 0, for example, while using Pin 1 to light up an LED or sound an alarm when the temperature rises above a certain level. You'll learn more about this in Chapter 10, 'Building Circuits'.

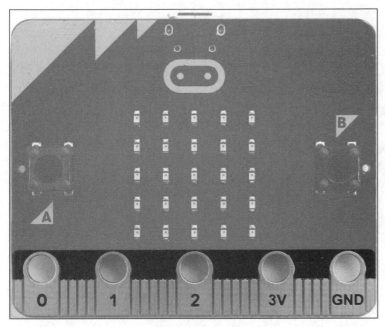

FIGURE 1-9: The BBC micro:bit's input-output pins

As well as the five large pins, the BBC micro:bit has a further 20 smaller pins. Unlike the large pins, these don't have holes for crocodile clips or banana plugs. Instead, using these requires an *edge connector*, as demonstrated in Chapter 11, 'Extending the BBC micro:bit'. Most micro:bit projects only need the three main input-output pins; the remaining pins are provided for more complicated projects.

Micro-USB Port

The BBC micro:bit's *micro-USB port* can be found to the middle of the upper rear edge (see Figure 1-10) and serves two purposes: It provides power to the BBC micro:bit when you're not using a battery pack, and it allows you to connect the BBC micro:bit to your computer to change its program and communicate data back and forth. These uses are explained in full starting in Chapter 2.

When using the micro-USB port, it's important to be careful and gentle. The inner connector is fragile, and if you damage it—such as by forcing a micro-USB connector in upside down or forcing a mini-USB, Lighting, or USB Type C connector into the micro-USB port—you won't be able to connect the BBC micro:bit to your PC any more. The port should also not be used to connect the BBC micro:bit to high-speed USB chargers, special USB charging ports, or high-power USB battery packs, which can potentially damage the BBC micro:bit. More information on this can be found on the official safety advice notice at `microbit.org/ guide/safety-advice`.

FIGURE 1-10: The BBC micro:bit's USB port

The micro-USB connector is commonly used in smartphones and tablets, and if you have a micro-USB cable for these, it should work fine for the BBC micro:bit, too. The only exception is in *charging-only cables*, often sold with power supplies and USB batteries, which don't have their data connections in place. If you connect the BBC micro:bit to your PC and it doesn't seem to be working, try a different micro-USB cable.

Battery Connector

The BBC micro:bit's small size, built-in display and sensors, and compatibility with conductive thread make it a great choice for portable and wearable projects, but having to power it from your PC or laptop is less convenient. That's where the *battery connector* comes in. This connector, on the top-right of the rear of the BBC micro:bit (see Figure 1-11), lets you connect a 3V battery pack to the BBC micro:bit and power your project on-the-go.

The connector is technically known as a *JST connector*. Not all batteries with a JST connector are suitable for use with the BBC micro:bit. Make sure that any battery pack you buy is listed as compatible and has been properly tested with the BBC micro:bit. These battery packs should only ever be used with standard disposable alkaline batteries. Rechargeable batteries use a lower voltage—1.2V rather than 1.5V—meaning the BBC micro:bit may not be able to get enough power and may be damaged.

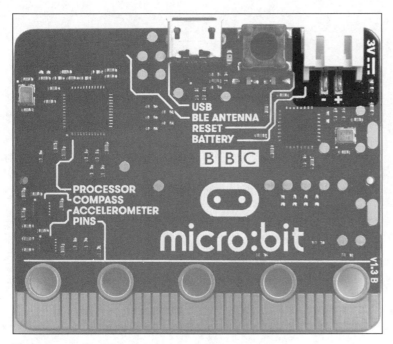

FIGURE 1-11: The BBC micro:bit's battery connector

A battery pack is provided with the BBC micro:bit Go bundle and with many third-party BBC micro:bit kits. If buying a battery pack separately to a BBC micro:bit, always check with the manufacturer or supplier that it is compatible; connecting a battery pack of the wrong voltage or polarity could damage your BBC micro:bit.

Chapter 2

Getting Started with the BBC micro:bit

In this chapter

- How to handle your BBC micro:bit safely and protect it against damage
- How to power your BBC micro:bit via USB or battery
- Exploring the BBC micro:bit's introductory program
- An overview of the BBC micro:bit's various inputs and outputs

HAVING EXPLORED YOUR BBC micro:bit's hardware in Chapter 1, 'Meet the BBC micro:bit', it's time to get down to the fun part: actually plugging it in and making it do things. Your BBC micro:bit can take its power from either your computer or a battery pack, and it comes complete with a preloaded sample application to help you get to grips with its various capabilities.

If you've already been using the BBC micro:bit and want to get started writing your own programs rather than using someone else's, feel free to jump ahead to Chapter 3, 'Programming the BBC micro:bit', for an introduction to loading new applications onto its memory.

Handling the BBC micro:bit

The BBC micro:bit is designed to be safe to handle and use without any kind of case over its circuitry. Because the BBC micro:bit operates on very low voltages, there's never a risk it will give you a shock; the same can't be said for the other way around, however.

All electronics are sensitive to what is known as *electrostatic discharge* (ESD). If you've ever walked on a thick carpet on a humid day and touched a metal door knob, you have experienced ESD for yourself in the form of an audible 'crack' and a bright spark arcing from your hand to the metal surface. ESD is a minor annoyance to humans, but to sensitive electronics it can be fatal.

To make sure that your BBC micro:bit remains safe from damage, take a few simple precautions against ESD. Start by always touching a metal surface before picking up the BBC micro:bit so that if you have built up a static charge it discharges into something other than the BBC micro:bit. When holding the BBC micro:bit, try to always hold it by the sides rather than by the copper-coloured strip along the bottom, and keep your fingers away from the small black components on the back of the board.

Powering the BBC micro:bit

Like any electronic device, the BBC micro:bit needs electricity to operate. It may be made of some clever components, but it still needs a power source before any of them will do anything interesting.

The most common way to power a BBC micro:bit is via its micro-USB port, by connecting a USB cable to your computer. Alternatively, you can insert two 1.5 V AAA or AA batteries into a battery holder which can be connected to the BBC micro:bit's secondary power input.

WARNING   The BBC micro:bit can be powered a third way: by connecting any 3V power supply to the pins marked '3V' and 'GND' on its edge connector. Doing so, however, bypasses protections against short-circuiting and supplying too high a voltage to the BBC micro:bit, and can result in damage to the BBC micro:bit if your clips shift during use. For more information on using the 3V pin, see Chapter 10, 'Building Circuits'.

USB Power

To power the BBC micro:bit from USB, you'll need a micro-USB cable. Some BBC micro:bit hardware bundles, such as the official BBC micro:bit Go bundle, include a cable, whereas others require you to provide your own. If choosing your own cable, make sure it is of a good quality and is not a 'charge-only' cable; these cables are only capable of providing power to a connected device and not making a data connection, and while they will power the BBC micro:bit you won't be able to use them to program it or communicate with it in any way.

Insert the smaller micro-USB end of the cable into the BBC micro:bit's micro-USB socket, taking care to ensure that it's the right way up. The curved side of the micro-USB connector faces down toward the table, and the BBC micro:bit should be sat with the LED display on the underside as in Figure 2-1. Take care not to force the cable; if the cable isn't going in with gentle pressure, take it out to realign and try again. When you've got it right, it should slot home with a gentle click.

FIGURE 2-1: Inserting micro-USB power

When the micro-USB cable is securely inserted, flip the BBC micro:bit over and insert the cable's other end into a free USB port on your computer or other USB power source (see Figure 2-2). The small yellow system LED on the back of the BBC micro:bit, between the micro-USB port and the Reset button, will light up when the BBC micro:bit is receiving power via the micro-USB port. Note that this only applies to USB power; if you're using a battery pack with the BBC micro:bit, this LED will not light.

If the BBC micro:bit has not already been loaded with a new program, it will begin running through the sample program detailed later in this chapter under the section 'Greetings from the BBC micro:bit'; otherwise, it will automatically run the last program from its memory. To switch the BBC micro:bit off, simply disconnect the micro-USB cable at either end.

FIGURE 2-2: A BBC micro:bit powered via USB

Battery Power

Powering the BBC micro:bit from a USB power source is quick and easy, but it's not very flexible. If you're planning on making use of the BBC micro:bit away from your desk, a better choice is to power the board using batteries installed in a protective holder, which then connects to the BBC micro:bit's battery header via a short length of wire. As with the micro-USB cable, some BBC micro:bit hardware bundles include a suitable battery holder. If yours did not, look for a holder capable of containing two AAA batteries, fitted with a JST 'PH' connector and specifically noted to be compatible with the BBC micro:bit.

WARNING The BBC micro:bit is designed to be powered via alkaline or carbon-zinc batteries only. Rechargeable batteries should not be used because they do not provide the right voltage. Whereas a nonrechargeable AAA battery outputs 1.5 V for a total of 3 V from two, for example, rechargeable AAA batteries output only 1.2 V for a total of 2.4 V. The BBC micro:bit may seem to work at this lower voltage; however, as the batteries discharge, it may begin behaving oddly and may lack the power to run external hardware (as described in Chapter 10, 'Building Circuits', and Chapter 11, 'Extending the BBC micro:bit').

To begin, install batteries in the battery holder. Most battery holders have a cover which is removed by pushing in the direction marked with an arrow; others require a small retaining screw to be undone before the cover can be removed as a means of preventing younger children from gaining access to the batteries.

Make sure that you insert the batteries correctly: the positive ends, marked with a + symbol and with a raised nub in the centre, should be toward the flatter parts of the battery connectors; the negative ends, marked with a – symbol and without the raised nub, should be against the battery holder's springs (see Figure 2-3).

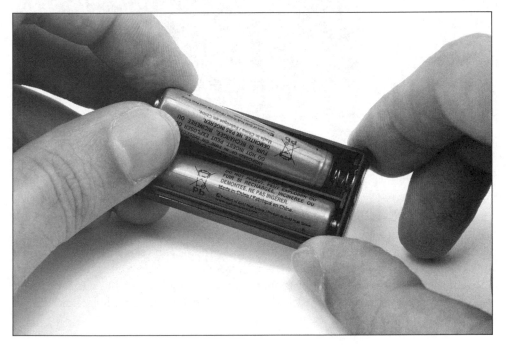

FIGURE 2-3: Inserting batteries

The JST 'PH' connector at the end of the battery holder's wire is *keyed*, meaning it can only be inserted in one direction. This key takes the form of a raised spike on the upper edge of the connector, which slides into a groove cut into the BBC micro:bit's battery connector. With the BBC micro:bit's LED display face down on the table, align the spine with the groove, check that the black wire is on the left side and the red wire to the right, and then push the connector home (see Figure 2-4).

FIGURE 2-4: Inserting the battery holder's PH connector

The BBC micro:bit has no power button, so as soon as the batteries are connected it will automatically power on (see Figure 2-5). As with USB power, if this is the first time the BBC micro:bit has been powered on, it will begin running through the sample program detailed later in this chapter under the section 'Greetings from the BBC micro:bit'; otherwise, it will automatically load and run the last program from its memory. To switch the BBC micro:bit off, simply disconnect the battery holder's JST connector; if your battery holder includes an on-board power switch, you can use this instead.

If the BBC micro:bit doesn't power on, check that the batteries are inserted fully into the holder and are the right way around. If it still doesn't work, the batteries may be too low in power; replace them with fresh batteries and try again. The yellow system LED, between the micro-USB connector and Reset button on the back of the BBC micro:bit, BBC micro:bit won't light up when running under battery power, so if you have loaded a program which doesn't use the LED display, the BBC micro:bit may be powered on even if it looks like it's off.

FIGURE 2-5: Powering the BBC micro:bit from a battery holder

Greetings from the BBC micro:bit

Fresh from the box, the BBC micro:bit comes preloaded with a sample program which walks you through its various capabilities, including using Button A and B and the accelerometer. This runs whether the BBC micro:bit is powered via USB or battery, and it acts as a great introduction to the platform.

If your BBC micro:bit has been used before and has been loaded with a different program, the new program will run instead of the factory-loaded sample program. If it does, you can skip the remainder of this chapter and move on to Chapter 3. If you or someone else has replaced the sample program and you want to load it back on again, you can download a copy from `support.microbit.org` by searching for 'first experience demo program'. You will learn how to load this onto the BBC micro:bit in Chapter 3.

The program demonstrates the capabilities of the BBC micro:bit's LED display, tests that both buttons are working, and shows how you can use the accelerometer as a more complex input. At the end of its run, which is detailed in full later in this section, it displays an attractive animation; if you'd like to start the program again, you can just press the Reset button on the back of the BBC micro:bit.

The sample program doesn't store information between uses. Every time the BBC micro:bit is powered on, the sample program starts again from the beginning. When you load your own program onto the BBC micro:bit, detailed in Chapter 3, that program replaces the sample program in the BBC micro:bit's memory.

If you'd like to go into the BBC micro:bit's introductory program with no expectations, stop reading at this paragraph and come back when you've run through the program at least once. To begin, power up the BBC micro:bit via USB or battery.

Signs of Life

The first thing you'll see when the BBC micro:bit runs through the sample program is an animation demonstrating the capabilities of the LED display. Although a 5 × 5 array of single-colour LEDs may sound limited, in reality it's enough to display some attractive graphics and animations; it can even be used to play a game, as you'll see later on in the program.

Following the animation, the BBC micro:bit will greet you with a simple word scrolling across the display: 'HELLO'. This is in reference to the classic first program written when learning a new programming language: 'Hello, World!'. You'll learn to write your own version of 'Hello, World!' in Section II, 'Coding for the BBC micro:bit', but for now just wait for the message to scroll and the program to move on to its next segment.

Testing the Buttons

With the animation and greeting over, the program moves on to its first interactive section. You'll see a flashing arrow pointing toward Button A, replaced at intervals with the letter A to further drive home the point: you should press Button A, on the left of the LED display, now.

When you press Button A, you'll be rewarded with another animation, followed by an arrow pointing toward the Button B replaced at intervals with the letter B. Press the Button B as instructed to fire off another animation, and the program will load its third segment: motion control.

Motion Gaming

The BBC micro:bit's accelerometer, detailed in Chapter 1, allows it to detect movement as well as the angle at which it is being held across all three dimensions. This is initially demonstrated in a simple manner: the message 'SHAKE!' will scroll across the LED display, followed by a simple animation. Shake the BBC micro:bit—being careful not to throw it across the room or rip out its power cable—and the program will proceed.

With the shaking over, the program will display the message 'CHASE THE DOT!' before playing a game where the aim is to—as you've probably guessed—chase a dot. The dot takes the form of an LED at the top right of the display, while your character is another LED in the middle of the display.

This game isn't played with the buttons, though. Holding the BBC micro:bit flat, begin to tilt the BBC micro:bit as though you're rolling a marble on a tray. You'll see the dot representing your character begin to move, as the accelerometer detects the changes in angle. Keep tilting until your dot lands on the target dot, and a new target dot will light up. When you've caught that one, too, the program will move into its final segment.

Get Coding

Having caught the dot, you'll see the encouraging message 'GREAT!' scroll across the screen, followed by a simple instruction: 'NOW GET CODING!'

At this point, the program is complete. The final segment loads a simple animation based around a picture of a heart, which will repeat over and over again. If you want to run through the program again, you can press the Reset button on the back of the BBC micro:bit to go back to the beginning; alternatively, if you disconnect it from its power source and reconnect it again, the program will begin from the start.

Although it may seem simple—and, really, it is—the sample program covers many of the BBC micro:bit's major features: using the LED display for pictures and animations; reading the buttons as inputs; using the accelerometer to detect gross motion and fine angles; and scrolling text across the LED display. These are the features you'll be using yourself as you learn to write your own code for the BBC micro:bit, starting in Chapter 3.

Resetting the BBC micro:bit

Whether you're using the sample program or one of your own making, it pays to become familiar with resetting the BBC micro:bit using the Reset button on the rear between the micro-USB and battery connectors. Pressing this sends a signal to the BBC micro:bit's processor to act as though the power has been briefly cut, starting the program again from the beginning.

The reset button (see Figure 2-6) is important, as there are a number of reasons you may want to interrupt a program's run and start again. There may be a bug in a program which has caused an *infinite loop*, whereby you can't progress any further. Some programs, like the preloaded sample program detailed earlier in this chapter, are only designed to run once and then stop entirely with the only way to restart being to reset the BBC micro:bit. When you're writing your own programs, you may want to test a program over and over again to iron out any bugs.

FIGURE 2-6: The BBC micro:bit's reset button

Regardless of why you want to start the program again, get used to using the reset button. While it's technically possible to restart a program by disconnecting and reconnecting power, this puts unnecessary strain on the power connectors and can cause damage to the BBC micro:bit over time.

You'll never damage the BBC micro:bit using the reset button. Unlike a desktop or laptop computer, which must go through a *shut-down process* before being powered off, the BBC micro:bit can be reset or powered off at any time without damage. Likewise, you can disconnect the BBC micro:bit's battery or USB cable at almost any time without fear of damaging it. The only exception is when you're loading a new program onto the BBC micro:bit, as described in Chapter 3. Although you won't cause any permanent damage by removing the cable before the process has finished, you may end up with a BBC micro:bit that doesn't respond properly. If this happens, simple reconnect it to your computer and try loading the program again.

Chapter 3

Programming the BBC micro:bit

In this chapter

- How to connect the BBC micro:bit to your computer to receive new programs
- Using the browser-based code editor to write your own programs
- An introduction to the BBC micro:bit simulator
- The difference between flash memory and RAM

PLAYING WITH THE BBC micro:bit's preloaded introductory program as introduced in Chapter 2, 'Getting Started with the BBC micro:bit', is a great way to learn your way around the hardware. To start getting the most out of the BBC micro:bit, though, you're going to want to replace the sample program with some code of your own.

In this chapter you will learn how to connect your BBC micro:bit to a computer so you can load new programs, look at the code editor in which you'll write your own programs, and learn a little bit about the difference between the BBC micro:bit's flash memory and its RAM. You'll also be given an overview of the languages in which you can program the BBC micro:bit, with more detail found in Chapter 4, 'Programming Languages'.

USB Connectivity

The micro-USB connector on the top edge of the BBC micro:bit serves two purposes. In Chapter 2, you learned how to power the BBC micro:bit through this connector, but it is also used to send data to and from a computer—whether that data is information gathered from the sensors or a brand-new program you've written.

To connect the BBC micro:bit to your computer, you'll need a micro-USB cable. If you didn't get one with your BBC micro:bit, you can use almost any good-quality cable. A cable that came with a tablet or smartphone will usually work, as will a low-cost cable purchased online or from a high-street shop. The only exception is so-called 'charge only' cables, designed exclusively for use with chargers and battery packs. These lack wires going to the data pins on the USB port. You'll know if you have one when you connect your BBC micro:bit to your computer and it seems to power on, but you don't see a removable drive (named MICROBIT) appear, as described later in this chapter. If this happens, you'll need to find a different cable with the data pins properly wired. You'll also need a spare USB Type A port on your computer. On a laptop, these ports are usually found at the side (see Figure 3-1); on a desktop computer, a small number of ports are usually found at the front, while a larger number can be found at the back amongst the other cables.

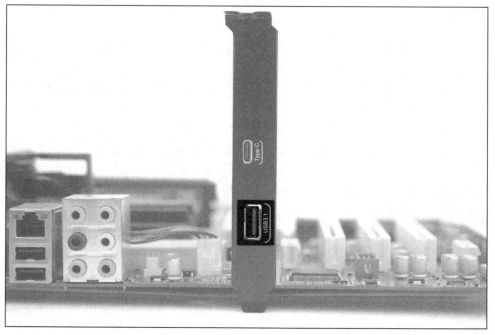

FIGURE 3-1: A USB Type A port

If you have a computer with only USB Type C ports available (see Figure 3-2), you'll need an adaptor which converts the USB Type C port to the more common USB Type A standard. Such adaptors are commonly supplied with the computer, but if not a search for 'USB Type C to Type A' at an online store will provide low-cost options that work fine with the BBC micro:bit.

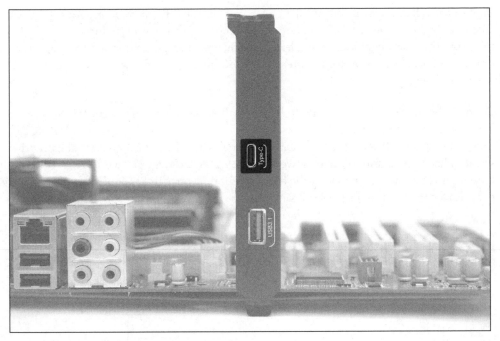

FIGURE 3-2: A USB Type C port

Drag-and-Drop

Unlike some development boards, the BBC micro:bit doesn't require you to install any additional software or drivers on your computer in order to load new programs. Instead, it appears as a *USB Mass Storage Device*, in the same way as an external hard drive or flash memory stick. When you first connect it, don't be surprised to see it running its sample program; the BBC micro:bit will continue to run whatever program was last loaded onto its memory whenever it has power, even as you're preparing to load a new program of your own devising.

Loading a program onto the BBC micro:bit is known as *flashing*, after the type of memory the BBC micro:bit uses for its storage: flash memory. More information about flash memory can be found at the end of this chapter, but for now all you need to know is that when the BBC micro:bit appears in the same way as a storage device on your computer; it's the flash memory you're accessing.

Traditionally, flashing a microcontroller on a board like the BBC micro:bit requires specialist software that has to be installed on every computer on which you want to program. The BBC micro:bit, though, is significantly simpler: when you have a program you want to load, simply drag it from its folder on your computer to the BBC micro:bit, and it will automatically load itself into memory.

Connect the micro-USB end of your cable to your BBC micro:bit and the larger USB Type A end of your cable to your computer now. After a few seconds, it will appear as a MICROBIT drive (see Figure 3-3). At a glance it looks the same as any flash stick or external hard drive you may have used before and shows two files: DETAILS.TXT and MICROBIT.HTM. The first file contains technical details about the firmware running on your BBC micro:bit; the second provides a handy link to the micro:bit website.

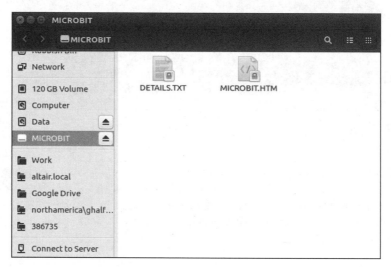

FIGURE 3-3: The BBC micro:bit as a removable drive

If you try to delete or edit either of these files, you'll find they're protected against modification. This is because the BBC micro:bit isn't a true removable disk like a flash drive; instead, you're seeing an interface between your computer and the firmware. If you try to copy a file from your computer—a picture, say, or a word processor document—to the BBC micro:bit, it will have disappeared the next time you plug it in without making the slightest bit of difference to the way the BBC micro:bit works.

The only files which the BBC micro:bit will not ignore when copied to the MICROBIT drive come in the form of programs you've written, which are read by the BBC micro:bit and stored in flash memory to replace the currently-loaded program. These programs aren't like the programs you'd run on your computer in that they're not *executable programs*; instead, they're *hexadecimal code*, which tells the BBC micro:bit what each block of its flash memory should store. These programs have the file extension `.hex` and are known as *hex files*.

Whether you're writing your own program or using someone else's, these hex files are downloaded using your computer's web browser. When you've downloaded a file, open your Downloads folder, click and hold the left mouse button while the cursor is over the file, and then drag it to the MICROBIT drive (see Figure 3-4). The BBC micro:bit will see the file, analyse it, and load it into memory before restarting so the new program can run.

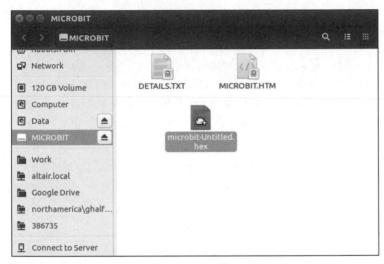

FIGURE 3-4: Dragging a hex file to the BBC micro:bit

When a new program is loaded onto the MICROBIT drive, the BBC micro:bit will automatically reset in order to begin running it. This will typically trigger a warning from your computer that what it is treating as a removable storage device has been removed without using the 'eject' or 'remove hardware' function. For a traditional removable storage device, removing it from the computer without using your operating system's 'eject' function can result in data being lost. The BBC micro:bit, however, is designed in such a way that it never needs to be ejected. If you see a warning message telling you otherwise, it can safely be dismissed.

Automatic Flashing

If you're running Microsoft Windows 7 or above, there's another way to flash new programs onto the BBC micro:bit: the micro:bit Uploader tool. Available from `makecode.microbit.org/uploader`, the tool runs in the background and monitors your Downloads folder for new hex files. When a new hex file is detected, it is automatically flashed onto your BBC micro:bit.

Be careful if you're using more than one BBC micro:bit on a single computer: the Uploader will flash new programs to every BBC micro:bit currently connected at the time you download the hex file. If you had a different program on any of the BBC micro:bits, this program will be overwritten.

To load a previously-downloaded program without downloading it again, simply use the drag-and-drop method outlined earlier in this chapter.

The Code Editor

Computer programs start their lives as simple text files with instructions on exactly what the computer should do. Although it's entirely possible to write a program in a text editor like Microsoft Notepad or a word processor like LibreOffice Writer, it's more common to use a tool known as an *integrated development environment* (IDE). At its heart, an IDE is a text editor designed specifically for programming and contains all the tools you'll need to code, debug, and compile your own programs.

With the BBC micro:bit, there's no need to install any IDE software on your computer itself. Instead, the IDE, known as the *code editor*, runs entirely in your web browser. Although the BBC micro:bit supports a range of different programming languages, the most commonly used is the JavaScript Blocks Editor, powered by Microsoft MakeCode.

Begin by loading the JavaScript Blocks Editor into your web browser by visiting `makecode.microbit.org`. You'll see a simple sample program already loaded (see Figure 3-5). When you first load it, the JavaScript Blocks Editor defaults to *Blocks* mode; clicking on the Editor toggle on the menu at the top of the screen will show you the same program in a different programming language, JavaScript. You'll learn more about the different languages in Chapter 4.

You will also see a picture of a BBC micro:bit, normally located on the left side of the screen. This is a key feature of the JavaScript Blocks Editor IDE. It's a *simulated* BBC micro:bit, which offers you a key insight into what your program will do even before you've loaded it onto the BBC micro:bit. Using this, you can try out changes to the program without having to take the time to test it on the BBC micro:bit hardware each time, or even if you don't have a physical BBC micro:bit available. The simulated BBC micro:bit is fully interactive: if you write a program which needs you to press Button A or B, just click on the picture's Button A or B to simulate the input.

If having the simulator running through a half-finished program is distracting you, simply click the Stop button underneath it (see Figure 3-6). When you want to test your program again, just click the button again to begin the simulator afresh.

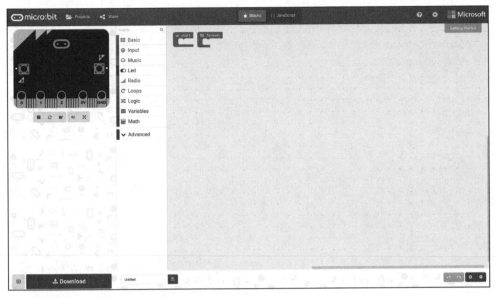

FIGURE 3-5: The JavaScript Blocks Editor

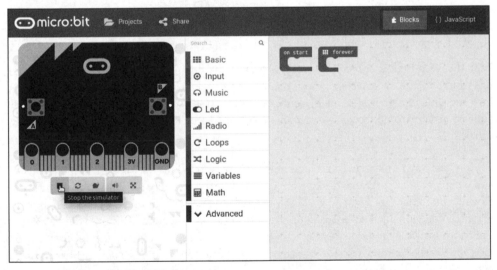

FIGURE 3-6: Stopping the simulator

Downloading Your Program

The sample program already loaded into the JavaScript Blocks Editor IDE is a great way to practise loading new software onto your BBC micro:bit using the drag-and-drop method you learned earlier in this chapter.

Without making any changes to the program, click the Download button (see Figure 3-7). The hex file for the sample program will automatically download onto your computer, while an informational screen offering step-by-step instructions on loading the file onto the BBC micro:bit will appear in your browser window.

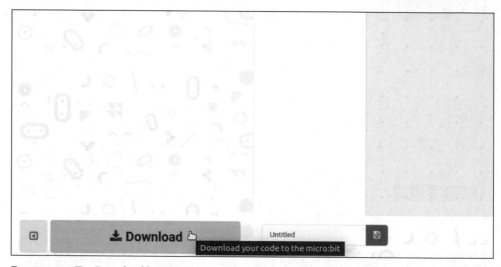

FIGURE 3-7: The Download button

Each download takes only a few seconds because the hex files are extremely small. If your BBC micro:bit isn't connected to your computer yet, take the micro-USB cable and insert it into the BBC micro:bit and one of your computer's free USB ports. The BBC micro:bit will appear on your computer as a removable disk called MICROBIT, ready to accept the new program.

In your computer's file manager—Explorer for Microsoft Windows, Finder for Apple macOS, and whatever your particular desktop environment offers if you're running Linux—find your Downloads folder (see Figure 3-8). Scroll through the list of files until you find the hex file you downloaded, called `microbit-Untitled.hex`—or, if you're using an older version of the Safari browser on Apple macOS, simply `Unknown`. Users of older Safari releases should look for the file `Unknown` wherever `microbit-Untitled.hex` or any other hex file name is referenced in this book—or upgrade to Safari 10.1 or newer to benefit from more intelligible file naming. Users of Safari 10 or older may find that the BBC micro:bit takes a long time to flash their programs. If so, rename the file to include the `.hex` extension before dragging it to the MICROBIT removable disk.

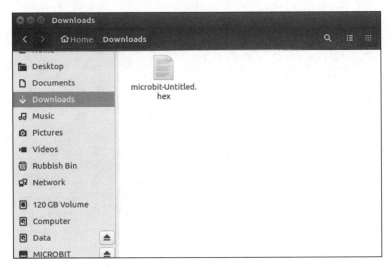

FIGURE 3-8: The Downloads folder

Click and hold the mouse button while the cursor is over `microbit-Untitled.hex`. Then, while still holding the mouse button down, drag the file across to the MICROBIT drive in the folder list (see Figure 3-9). Release the mouse button, and the file will be copied across to the BBC micro:bit's flash memory and automatically loaded. You can see this happening by looking at the yellow system LED between the micro-USB port and the Reset button: the LED will flash as the program is flashed, then go back to being steadily lit once the program has finished loading. Although the short starter program you've just flashed is valid, it doesn't contain instructions telling the BBC micro:bit what to do, so it simply won't do anything at all.

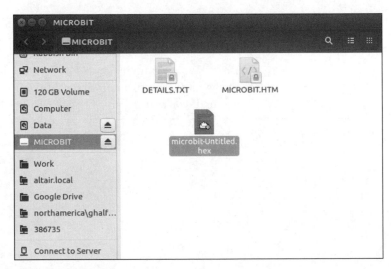

FIGURE 3-9: Flashing the example program

Congratulations! You've successfully loaded your first program onto the BBC micro:bit. If you're having difficulty finding the MICROBIT in your file manager, Figures 3-10, 3-11, and 3-12 will help.

WARNING If you've connected your BBC micro:bit to your computer and still don't see a MICROBIT drive, try switching the micro-USB cable to a different USB port on your computer and restarting your operating system.

If you still can't see a MICROBIT drive, try a different micro-USB cable. Some cables, especially those sold with USB power packs and similar devices, only have the wires they need to provide power and not data signals; these cables will work to power the BBC micro:bit but can't be used to flash new programs or communicate with the BBC micro:bit. If you're sure you're using a fully-wired micro-USB cable and still can't see the MICROBIT drive, visit `support. microbit.org` for further troubleshooting steps.

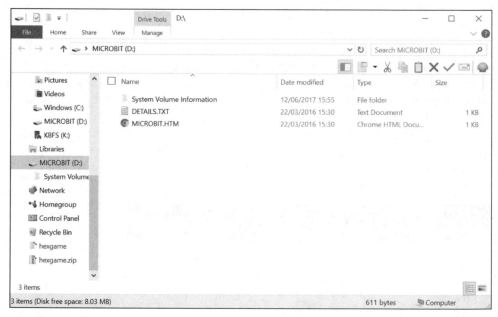

FIGURE 3-10: The MICROBIT drive on Windows 10

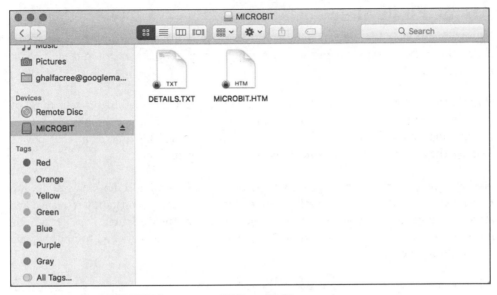

FIGURE 3-11: The MICROBIT drive on macOS Sierra 10.12

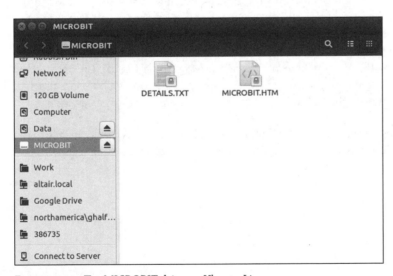

FIGURE 3-12: The MICROBIT drive on Ubuntu Linux

About Flash Memory

To understand how the BBC micro:bit stores its programs, it's necessary to know a little about the different types of memory a computer has: *volatile memory* and *non-volatile memory*.

Volatile memory is a temporary storage area, akin to keeping something in your mind by repeating it over and over again under your breath. It needs constant refreshing and stores information only as long as the computer—or, in this case, BBC micro:bit—has power. Remove the power, and the contents of the memory are gone forever.

Non-volatile memory, by contrast, is more permanent, like writing something you need to remember in a notebook. Non-volatile memory doesn't need to be constantly refreshed and stores its contents even when power is disconnected from the computer. Just like a notebook, however, non-volatile memory is slower both to store new data and to find the data you've already stored.

When you flash a program onto the BBC micro:bit, you're loading it into a type of non-volatile storage called *flash memory*. Also known as *program memory*, for obvious reasons, anything stored in this area is available to the BBC micro:bit whenever it is powered on; that's why when you disconnect the BBC micro:bit from your computer and reconnect it in the future, it'll automatically run the last program you had loaded. It's also why we call the process of loading a program onto the BBC micro:bit *flashing*. When the program is running, it is using a type of volatile memory known as *random access memory* (RAM). This means that anything you do in your program, such as counting the number of times a button is pressed, will be forgotten if you reset the BBC micro:bit or remove and reinsert its power cable.

The process of flashing a program onto the BBC micro:bit is as simple as dragging a file to the MICROBIT drive, as demonstrated earlier in this chapter—no matter how scary the phrase may sound!

Part II

Coding for the BBC micro:bit

Chapter 4
Programming Languages

In this chapter

- An overview of programming languages and their necessity
- A look at the three main programming languages for the BBC micro:bit: JavaScript Blocks, JavaScript, and MicroPython
- A guide to choosing a programming language

TO GET THE most from your BBC micro:bit, you're going to have to learn to write your own programs. Before you can jump into programming the device, however, you have a decision to make: which of the three main programming languages—JavaScript Blocks, JavaScript, and Python—to use.

About Programming Languages

Computers are surprisingly simple devices. At its heart, a computer is little more than a series of switches which can be either on or off. While your house might have a dozen switches for lights, a computer has billions upon billions of switches which can do incredible things when combined in interesting ways.

Before a computer can do anything, though, it needs to be programmed. When you turned on your BBC micro:bit and ran through the introductory program in Chapter 2, 'Getting Started with the BBC micro:bit', you were running a program someone had previously written and loaded onto your BBC micro:bit. Without that program, there would have been no smart little animation or friendly messages; the BBC micro:bit would have simply done nothing.

At its lowest level, a computer's processor is given a series of instructions in *machine code*. This is a symbolic language filled with short instructions like ADD, MOV, and BNE, and it works on locations in the processor's memory called *registers*. While it's entirely possible to write complex programs in machine code, it can get confusing, which is why higher-level *programming languages* were invented.

These languages sit at a level above machine code. They are much easier for people to understand and write than machine code, using friendly instructions like 'print' to do what might take a machine code program dozens of instructions to complete. They're harder for the computer to understand, though, and must be translated into machine code before being run—like how a document written in French would need to be translated before it would be accessible to someone who could only read English.

If each line of code in your program is run as-is on the BBC micro:bit, without needing to be translated into machine code first, the language is known as *interpreted* and is the equivalent of having an interpreter translating foreign speech for you as it happens. If the translation to machine code happens all at once when you have finished writing the program, it is known as a *compiled* language and is the equivalent of having an entire written document translated into your language before you start to read it.

Sometimes the lines between interpreted and compiled languages are blurred. The languages used on the BBC micro:bit can be either interpreted or compiled, but they are typically compiled once you've written them in an editor and then the compiled program is flashed onto the BBC micro:bit (see Chapter 3, 'Programming the BBC micro:bit'). In some instances, though, you might use the BBC micro:bit as an interpreter, as in Chapter 9, 'The BBC micro:bit and the Raspberry Pi'.

The Three Main BBC micro:bit Languages

There are hundreds of programming languages, many specialised for a particular purpose, such as distributing a complex program over many different processors or writing graphical programs and games. For the BBC micro:bit, though, there are three 'core' languages for which official browser-based editors are available:

- JavaScript Blocks
- JavaScript
- Python

JavaScript Blocks

The JavaScript Blocks language will likely be most people's first introduction to the BBC micro:bit. Based on the design of Scratch, which was originally developed by the Massachusetts Institute of Technology's Media Lab, JavaScript Blocks is a *visual programming language*. Rather than typing instructions by hand from memory or consulting a reference guide, all the instructions available to the language are available in a *Blocks toolbox*. Click on the block containing the instruction you need, drag it to the workspace, and connect it to other instructions to build your program (see Figure 4-1).

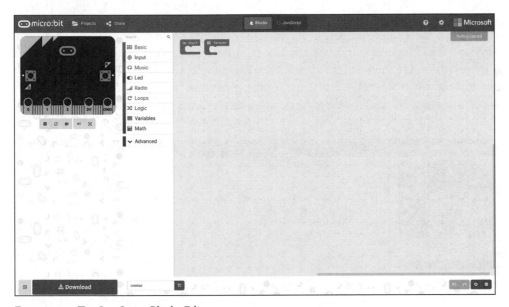

FIGURE 4-1: The JavaScript Blocks Editor

The visual nature of JavaScript blocks makes it an ideal first programming language for younger coders, while the way blocks click together like jigsaw pieces means it's also a great way to learn fundamental programming concepts. It's easy to see how a *loop*—a section of program which repeats depending on a particular input, such as whether or not you're holding down a button or a sensor has a particular reading—works in JavaScript Blocks, for instance, thanks to colour-coded blocks and the way the repeated blocks nest inside the control block.

If you've previously used Scratch, there's one important difference: the *stage*, the area of the screen onto which *sprite* objects can be placed and animated, doesn't exist in JavaScript Blocks. Scratch programs are designed to be run on your computer; JavaScript Blocks programs are designed to be run on the BBC micro:bit, and the BBC micro:bit itself forms the stage.

Although the idea of dragging and dropping blocks of code may seem basic, JavaScript Blocks is an extremely powerful language in which it's possible to make complex and substantial programs. It's also a gateway language. As you're writing a JavaScript Blocks program, the editor is quietly translating it into text-based JavaScript in the background. When your program is complete, you're able to see the JavaScript version with all the graphical block elements replaced with their raw instructions with the click of a mouse.

JavaScript

Originally released in 1995, JavaScript has become the language of the World Wide Web. Not to be confused with Java, an entirely separate programming language with which it confusingly shares the first part of its name, JavaScript is flexible, lightweight, and *cross-platform*. The latter means it can run on a wide variety of computers, from powerful servers all the way down to the BBC micro:bit. It's also a traditional text-based language, rather than the visual JavaScript Blocks. Instructions are represented as lines of writing rather than colourful blocks and are typically entered by typing them out (see Figure 4-2).

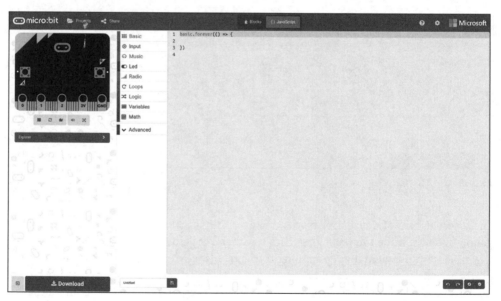

FIGURE 4-2: The JavaScript Editor

The JavaScript Editor is identical to the JavaScript Blocks Editor. It's possible, in fact, to switch between the two at will even part-way through writing a program. A program written in JavaScript Blocks can be viewed and edited in JavaScript proper, and at least partially the other way around. If you've used features available in JavaScript that aren't available in JavaScript Blocks, such as *functions*, they'll appear as grey blocks you can move and delete but not edit. Even one-way translation from JavaScript Blocks to JavaScript makes it the obvious choice for progressing from a visual programming language to a more traditional one.

Another advantage of the JavaScript Editor is that it retains the Blocks toolbox of JavaScript Blocks. Whereas clicking on an instruction block in JavaScript Blocks puts a block within the visual editor for you to click and drag around, clicking on the same instruction block in JavaScript inserts a copy of the instruction—complete with any gaps you might need to fill in, such as the numbers you want to compare or the message you want to print—on your current line.

Some features available in JavaScript are not available in JavaScript Blocks, however. When writing more complex programs, JavaScript gives you the ability to create subroutines and functions—a capability JavaScript Blocks presently lacks—although only by hand rather than by using the Blocks toolbox the language shares with JavaScript Blocks.

JavaScript was originally implemented as an interpreted language, but the BBC micro:bit uses it as a compiled language. When you click the Download button in the JavaScript (or JavaScript Blocks) Editor, your program is translated to machine code and compiled into a file the BBC micro:bit can run once flashed onto its memory.

Python

Created by Guido van Rossum and first released in 1991, Python is a hugely popular programming language for both education and professional development. Code written in Python can be more readable to the inexperienced than the same code written in JavaScript, thanks to the use of white space to indicate blocks of code rather than curly braces (see Figure 4-3). Technically, the Python available for the BBC micro:bit is variant Python designed specifically for use with microcontrollers and more properly called *MicroPython*, although the underlying language itself is still called Python.

Unlike JavaScript, there is currently no 'Blocks' variant of Python. Programs are entered into the editor the old-fashioned way, typed out by hand using a reference guide for the instructions—or, when you're experienced enough, entirely from memory—with a Snippets menu for inserting the most common code segments automatically. Although this can make the Python Editor seem forbidding compared to the JavaScript Editor, the language itself is designed to be accessible and easy to read and write with a minimum of programming experience.

FIGURE 4-3: The Python Editor

One of the major advantages to Python is its use in other educational programming projects. The Raspberry Pi Foundation, as an example, uses Python as one of the languages for its Raspberry Pi single-board computer; any school or coding club which uses Raspberry Pi will be familiar with the Python language, and that familiarity gives them a head start when using the same language on the BBC micro:bit, just as anyone using Python on the BBC micro:bit will find Python on the Raspberry Pi comfortably familiar.

As with JavaScript, Python was originally devised as an interpreted language but is used on the BBC micro:bit as the functional equivalent to a compiled language, in which the program is written in full then compiled into a hex file by the code editor, ready to be flashed onto the BBC micro:bit. The exception lies in a powerful feature of Python: *Repl*, an interactive interpreted *shell* which runs your commands as soon as you type them. Repl is explored in more detail in Chapter 9, where it is used to allow the Raspberry Pi to take control of the BBC micro:bit.

Comparing Programming Languages

Table 4.1 offers a quick comparison of each language's major features.

Table 4.1   Comparing the BBC micro:bit Languages

Feature	JavaScript Blocks	JavaScript	Python
Browser-Based Editor	Yes	Yes	Yes
Cross-Platform	Yes	Yes	Yes
Free	Yes	Yes	Yes
Learning Materials Available	Yes	Yes	Yes
Visual Environment	Yes	Yes	Yes
Integrated Instruction Reference	Yes	Yes	Partial
Official Micro:bit Simulator	Yes	Yes	No
Functions and Subroutines	Partial	Yes	Yes
Built-In Event Handling	Yes	Yes	No
Integrated Compiler	Yes	Yes	Yes
Interactive Shell	No	No	Yes

These features are explained next.

- **Browser-Based Editor**—Whether a language has an editor you can run in the web browser of any computer, without the need to install software on the computer.

- **Cross-Platform**—Whether a language's editor and compiler can be used on any computer, regardless of its operating system.

- **Free**—Whether it costs any money to be able to use a language, or whether the editor or compiler needs to be purchased before it can be used.

- **Learning Materials Available**—Whether a language has readily learning materials, both for personal and for classroom use.

- **Integrated Instruction Reference**—Whether a language has a list of the most common instructions available directly within the editor as a reference guide. A 'no' in this row does not indicate that guides are not available separately.

- **BBC micro:bit Simulator**—Whether a language has a BBC micro:bit simulator built into its editor, allowing you to quickly test your program without having to compile, download, and flash it onto your BBC micro:bit.

- **Functions and Subroutines**—Whether a language has the ability to create functions and subroutines, small programs-within-programs that can be called multiple times from the main program to reduce the complexity of the program and save time otherwise spent writing duplicated code.

- **Built-In Event Handling**—Whether a language has the ability to monitor for events happening outside the main program loop, such as button presses, and interrupt the flow of the program to run a separate piece of code when such events are triggered.

- **Integrated Compiler**—Whether a language has a compiler built into its editor or requires you to download the program code and compile it in a separate program.

- **Interactive Shell**—Whether a language has the ability to run on the BBC micro:bit as an interactive shell, interpreting your commands as you type them, regardless of whether that functionality is exposed in the language's browser-based code editor.

Choosing a Programming Language

A programming language is, for most, a personal choice. If you're using your BBC micro:bit at school or at a coding club, the choice may have already been made for you. Lesson plans, sample programs, and other materials based on one of the aforementioned languages will have been prepared for you, and to avoid confusion, it's best to use the same language when you're experimenting at home as well.

For younger or less experienced users, JavaScript Blocks is the obvious choice. Its highly-visual programming environment with colour-coded instruction blocks, jigsaw-like drag-and-drop functionality, and always-handy references mean it's as simple as possible to get started. It's also a great language for those who wish to minimise the amount of typing they have to do and helps to reduce basic errors like misspelling an instruction or forgetting to close a loop.

JavaScript itself, meanwhile, is the logical next step for anyone who has been honing their skills on JavaScript Blocks. The instructions are identical to JavaScript Blocks, just presented in a different format, making it familiar, while the presence of the Blocks toolbox in the JavaScript Editor provides a quick reference and acts as a shortcut to minimise typing.

Finally, Python should be the first choice for anyone who has already been working with the language on other platforms. The core concepts within Python for the BBC micro:bit are no different to Python running on a Raspberry Pi or a desktop or laptop computer, although the BBC micro:bit has a number of unique instructions for handling its sensors and screen. It's also a great choice if you are looking to link the BBC micro:bit up to another computer, allowing you to write both sides of the program in a single language, as demonstrated in Chapter 9.

The important thing to remember, though, is that all the languages are equally functional. Given enough time and experience, you can create the same program in any of the three languages, something you'll see demonstrated in the following language-specific chapters.

Other Programming Languages

The languages detailed in this chapter are the three most common, and the ones chosen by the Micro:bit Educational Foundation for official support. The open nature of the BBC micro:bit, though, means that it supports a variety of different languages and programming tools, including C/C++, Rust, Forth, Pascal, and Ada.

If you already program in a particular language, you may find that support for writing micro:bit programs already exists. Look for 'micro:bit' and the name of your language on your favourite search engine to find out if support is available.

Chapter 5
JavaScript Blocks

In this chapter

- An introduction to programming in the JavaScript Blocks Editor
- 'Hello, World!': your first JavaScript Blocks program
- Programs for reading from button inputs, touch inputs, and the temperature, accelerometer, and compass sensors
- Fruit Catcher: a simple game to program and play on your BBC micro:bit

JAVASCRIPT BLOCKS IS a great entry into programming for younger readers and those without previous computing experience. Its visual environment makes it easy to see which parts of a program go together, and by dragging prewritten code blocks from the toolbox, mistakes are far harder to make than in traditional nonvisual programming environments.

In this chapter you'll be introduced to the JavaScript Blocks Editor, powered by Microsoft MakeCode, and learn to write a series of programs that make the most of the BBC micro:bit's capabilities. Finally, you'll be able to make a quick-fire game in which you're given the task of catching ever-faster falling fruits before they hit the ground.

Introducing the JavaScript Blocks Editor

The JavaScript Blocks Editor is an all-in-one development environment created specifically for the JavaScript Blocks language. Running entirely within the web browser of any modern computer, the JavaScript Blocks Editor doesn't require you to install any software on your computer. All you need to do to load the JavaScript Blocks Editor is open the web browser of an Internet-connected computer, type **makecode.microbit.org** into the address bar, and press the Enter key (see Figure 5-1). Once the editor has loaded, it runs directly on your computer even if you disconnect from the Internet.

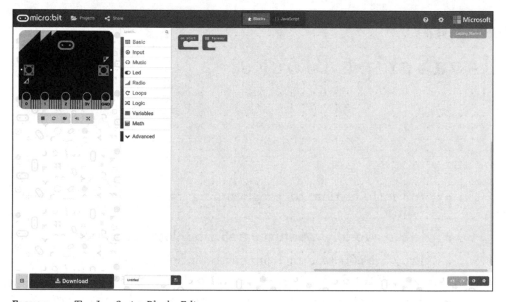

FIGURE 5-1: The JavaScript Blocks Editor

The main features of the JavaScript Blocks Editor follow:

- **The Projects menu**—Located at the top left and looking like a folder icon, use this menu to create a new, blank project or to load a previous project. These project files are stored on your computer, in your web browser's local storage area, so they are available only to you unless you choose to share them. Be aware that clearing stored data using your web browser's 'clear history' option will delete these files, so make sure to save any you wish to keep using the Save Icon. The Projects menu also offers access to sample projects under the Make tab, and code samples under the Code tab.

- **The Share button**—Located next to the Projects menu and shaped like a circle with two further circles attached on sticks, use this button to publish and share a program you've created in the JavaScript Blocks Editor.

- **The Editor toggle**—Looking like a jigsaw piece next to some lines of writing and located in the middle of the top menu bar, use this to toggle between JavaScript Blocks and JavaScript modes. The former mode is the default, using the visual editing environment; the latter mode is explained in full in Chapter 6, 'JavaScript'.

- **The Help button**—Found to the right of the top menu bar and shaped like a question mark in a circle, use this button to access the JavaScript Blocks Editor's built-in help, including a getting started tutorial—also available using the orange Getting Started button to the right—and sample projects.

- **The Settings button**—Located next to the Help button, the Settings button is shaped like a cog. Use this button to change project settings, add additional packages such as libraries for using extra hardware with your BBC micro:bit, delete a project, or delete all your saved projects. You can also view the Editor's privacy policy and terms of use and provide feedback from here.

- **The simulator**—Located to the left on a high-resolution screen or to the bottom left on a lower-resolution screen, the simulator is more than just a picture of a BBC micro:bit. When you write your program, it will automatically begin running in the simulator exactly like it would on the BBC micro:bit itself. This allows you to test your program without having to download the compiled hex file and flash it to your BBC micro:bit. The simulator is fully interactive: to simulate pressing Button A, just click the picture of Button A. If your program uses one of the sensors, the simulator will adjust to allow you to use those, too, adding a temperature slider for the thermometer or letting you change its angle for the accelerometer or compass.

- **The Blocks toolbox**—Located in a strip to the left of the screen, you will use the Blocks toolbox to build your program. Blocks of code—shaped like jigsaw pieces—are dragged out of the Blocks toolbox and onto the Workspace as you begin coding. Each block is separated into a series of categories: Basic, Input, Music, LED, Radio, Loops, Logic, Variables, and Math, with additional blocks available by clicking the Advanced drop-down menu.

- **The Workspace**—Taking up the bulk of the screen and decorated with plus symbols, you'll build your program on top of the Workspace. When you start a new project, you'll already have two blocks on the Workspace: one labelled [on start] and the other labelled [forever].

- **The Download button**—When you've finished your program and want to run it on your BBC micro:bit, just click the Download button at the bottom of the screen and the JavaScript Blocks Editor will compile it and download the hex file for you to drag and drop onto the MICROBIT drive (see Chapter 3, 'Programming the BBC micro:bit').

- **The project name**—The first thing to do when you start a new project is to give it a descriptive name. Always start by renaming a project from the default 'Untitled' to make it easier to find in the future.

- **The Save icon**—Located next to the Project Name, the Save icon allows you to save a copy of your program to your computer. If you want to be able to take a copy home from school or vice versa, the Save icon will let you do this. You can load saved projects into the JavaScript Blocks Editor by clicking on the Projects menu and then Import File.

- **The Undo and Redo buttons**—Shaped like curled arrows, the Undo and Redo buttons allow you to undo a mistake by clicking on the leftmost arrow and then redoing it by clicking the rightmost arrow—just in case undoing the mistake was the real mistake!

- **The Zoom buttons**—Found at the bottom right of the screen, use these buttons to zoom in and out of the Workspace to make the blocks larger or smaller. For more complex projects, zooming out allows you to see more of the project on-screen at once; if you find the writing on the blocks hard to read, zooming in will make things clearer.

If you want a walk-through of using the JavaScript Blocks Editor, click on the Help button followed by Getting Started or click the orange Getting Started button directly. When you're comfortable, proceed with the rest of the chapter to begin writing your own programs for the BBC micro:bit.

Program 1: 'Hello, World!'

'Hello, World!' is the traditional first program for any language. It's basic by design but offers a first glimpse of what a given language is like to program in with the bonus of a quick payoff in the form of a message proving that your program is alive and well.

Start by opening your browser and going to `makecode.microbit.org` to load the JavaScript Blocks Editor. If you've been working on projects previously, click the Projects menu and New to load a new, blank project. Click on the box labelled Untitled at the bottom of the screen to rename your program, and type **Hello World** to give it its new name.

'Hello, World!' is a simple program that requires only one extra block in addition to the 'on start' and 'forever' blocks you have on your screen. Click on the Basic category in the Blocks toolbox to the left of the Workspace, and then click on the `[show string "Hello!"]` block to place it into the Workspace (see Figure 5-2).

TIP If you make a mistake and drag the wrong block onto the Workspace, simply click the Undo button at the bottom right of the screen to send it back to the Blocks toolbox. Alternatively, right-click on the block and click the Delete Block option.

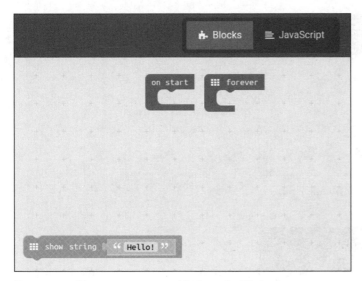

FIGURE 5-2: The [show string] block on the Workspace

You may have noticed that your new block is covered in diagonal lines. These are used to indicate that the block won't run because it isn't inside what is known as a *control block*. These control blocks sit around other blocks and control when and how they should run. Each block is shaped like a jigsaw piece and will only slot into particular other blocks. Control blocks have smooth outside edges and gaps in the middle, indicating they go around other blocks; the [show string "Hello!"] block you have put onto the Workspace has an indent on the top which fits the one inside the control block perfectly.

Drag the [show string "Hello!"] block into the [on start] block by clicking on it, holding down the mouse button, moving the cursor over the hollow inside of the [on start] block, and then letting go of the mouse button. The block should snap into place; if not, drag it again to adjust its position until it does. You'll know you've got it right when the block's diagonal lines disappear (see Figure 5-3).

At this point, the simulator will fire into life. By placing the [show string "Hello!"] block into the [on start] block, you've completed your first program. The message "Hello!" will scroll across the simulator's display, just like it would on a real BBC micro:bit—but that's not the message we want, so the program needs to change.

In the [show string "Hello!"] block, click on the word "Hello!" to highlight it. This section of the block can be changed as desired, so type out the message you actually want: **"Hello, World!"** The block will change accordingly, and what was a [show string "Hello!"] block will become a [show string "Hello, World!"] block (see Figure 5-4).

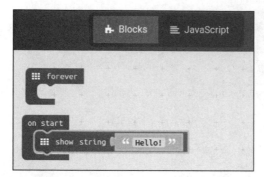

FIGURE 5-3: The [show string] block in place

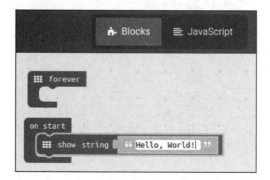

FIGURE 5-4: Changing the [show string] block message

When you change a block in place like this, the simulator will automatically restart the program, and you'll see the message "Hello, World!" scroll across the screen. Feel free to experiment at this point: change the block to scroll a message of your own, such as your name or your favourite colour.

TIP   In programming, messages like this are called *strings* and can be made up of letters, numbers, and some symbols. Although a string is great for displaying messages on a screen, you're limited in how you can work with it: the string "1", for example, can't be used in a mathematical equation because while it looks like a number to us it's simply a string to the programming language. For mathematical equations, there are other types: *integers* are whole numbers, and *Booleans* are logic elements which can be True or False—equivalent to 1 and 0, respectively.

Whatever message you decide to scroll, you'll notice it only scrolls once and then the simulator stops. This is the result of placing the [show string "Hello, World!"] block into the [on start] block. Whatever is in the [on start] block will only ever run once when the program begins. You can make the message appear again by resetting the simulator. You do this by pressing the icon shaped like two bent arrows underneath the picture of the BBC micro:bit, but if you don't want to have to keep doing that, you're going to need a method of telling the program to repeat.

Loops

Click and drag the [show string "Hello, World!"] block out of the [on start] block and into the [forever] block (see Figure 5-5). Watch the simulator as you do. The program will restart, as it always does when you make a change, but rather than scrolling the message once and then stopping, the message will repeatedly scroll forever.

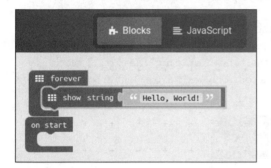

FIGURE 5-5: Looping the [show string] block message

In programming, this is known as a *loop*. When the program reaches the end of its instructions—in this case, when it has finished scrolling the message from the [show string "Hello, World!"] block—it returns to the beginning of the loop and starts all over again. As the name implies, the [forever] block will loop forever; in programming, this is known as an *infinite loop*.

If you want the program to stop, you can either remove the [show string "Hello, World!"] block from the [forever] block or click the square Stop button under the simulator.

The key thing to take away here is the difference between the [on start] block and the [forever] block. The [on start] block will only run through its contents once, when the program begins; the [forever] block will loop its contents forever.

TIP

If you want to see how the program runs on a real BBC micro:bit, click the Download button to compile it into a .hex file and drag it to the MICROBIT drive, as described in Chapter 3.

Program 2: Button Inputs

'Hello, World!' demonstrated how to get an output from the BBC micro:bit via the screen, but there's another key feature of most programs: inputs. Thankfully, the BBC micro:bit has two inputs that are ready to use: Button A and Button B, either side of the display, and now it's time to introduce those into your program.

To ensure you don't lose your 'Hello, World!' program, start a new project by clicking on the Projects menu and clicking on New. You'll get a fresh project with two control blocks (see Figure 5-1 earlier in this chapter), and any changes you make won't affect your previous project. To keep the two easy to find, remember to give this new project a name: click on Untitled at the bottom of the screen; then type **Button Inputs** to rename the project.

Begin by clicking on the Input category of the Blocks toolbox and then clicking on the [on button A pressed] block. When it appears on your Workspace, you'll notice that it doesn't have the diagonal lines you saw on the [show string "Hello!"] block. If you look closer, you'll notice that it's shaped like a control block, too (see Figure 5-6).

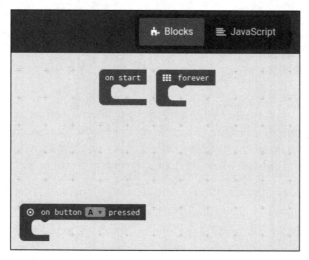

FIGURE 5-6: The [on button A pressed] block

The [on button A pressed] block is an *event block*, and it works a little differently from the blocks you've been using before. An event block sits outside the main program. Any blocks within do not normally run, but that changes when its *trigger* is activated. In the case of the [on button A pressed] block, that trigger is Button A being pressed. When that trigger is activated, the BBC micro:bit will jump into the [on button A pressed] block as soon as it is able, and the blocks within it will begin to run.

At the moment, there are no blocks inside the [on button A pressed] block. To fix that, click on the Basic category of the Blocks toolbox; then click on the [show icon] block with the picture of a heart on it. This block type tells the BBC micro:bit to show a picture from a selection already drawn for you, rather than scrolling a string as in 'Hello, World!'. Drag the [show icon] block into the [on button A pressed] block, and then click on the heart picture to bring up the list of alternative icons. Find the happy face icon (see Figure 5-7), and then click it to select it.

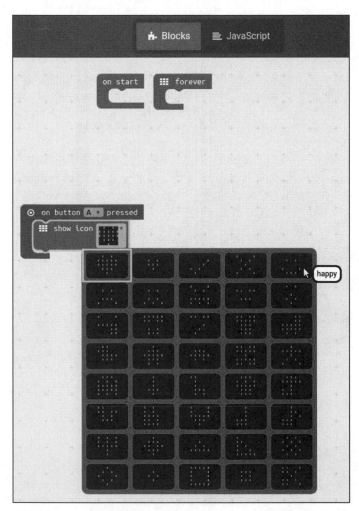

FIGURE 5-7: Choosing a happy face icon

If you click on Button A in the simulator now, you'll see the effect of your program: a happy face will appear on the screen. It'll stay there even when you're not holding the button down, too, because the program doesn't say otherwise. There's no block to display a different image or to clear the screen, so once the happy face is there, it stays there until you reset the simulator or BBC micro:bit.

Event blocks only run when their trigger is activated and will run each time the trigger is activated: press Button A once, and your Event Block will run once; press it again and it will run again; press it a third time, and it will run a third time, and so on.

Multiple Buttons

Being able to read a single button is great, but the BBC micro:bit has two. Thankfully, adding in the second button is as simple as repeating what you've already done. Click on the Input category of the Blocks toolbox, click on the [on button A pressed] block, click on the Input category of the Blocks toolbox, and then click on the [show icon] block.

Drag the [show icon] block into the [on button A pressed] block, and then click on the heart icon and choose the sad icon to replace it. Finally, click on the 'A' of the [on button A pressed] block and change it to 'B' to have it trigger when you press Button B instead of Button A (see Figure 5-8).

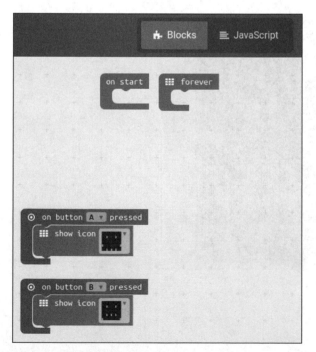

FIGURE 5-8: The finished multibutton program

Click on Button A in the simulator—or press the physical button if you're running the program on a real BBC micro:bit—and the happy face will appear as before. Click Button B, though, and you'll get a sad face. Your program can read either button and run different blocks depending on which is pressed.

There's a third option for reading buttons, too: reading both at the same time. If you click on either the [on button A pressed] block or [on button B pressed] block and change them again, you'll see the option A+B. An [on button A+B pressed] block will run only when both buttons are pressed at the same time, allowing you to have up to three inputs from two buttons.

As an experiment, try adding another [on button A pressed] block, changing it to an [on button A+B pressed] block, and picking a different icon to display. On a real BBC micro:bit, hold down both Button A and Button B to make the new block run. On the simulator, where you can't click on both buttons at the same time, you'll see an A+B button appear just below Button B. Pressing the A+B button is the equivalent of pressing both buttons at the same time on a physical BBC micro:bit.

Program 3: Touch Inputs

Although its two buttons give the BBC micro:bit up to three possible inputs, that might not be enough for your needs. This is where the input-output pins, the numbered contacts on the copper strip at the bottom of the BBC micro:bit, come into play. The BBC micro:bit can detect when you touch one of them with your finger and react accordingly—a fun alternative to the physical buttons and a great way to add extra inputs to a program without the need to buy and wire up more buttons.

As always, you need to start a new project by clicking on the Projects menu and clicking on New so you don't lose your work on the Button Inputs project. Click on Untitled at the bottom of the screen; then type **Touch Inputs** to name your new project.

Remember that you can always find your old projects by clicking on the Projects menu and looking under the My Stuff section. Projects are ordered according to when they were last modified, so your newest projects will be first and your oldest projects last. **TIP**

Reading a pin is a lot like reading a button. Click on the Input category of the Blocks toolbox and click on the [on pin P0 pressed] block. Like the [on button A pressed] block, the [on pin P0 pressed] block is an event block; it doesn't need to be inside the [on start] or [on forever] blocks, and any blocks placed within it won't run until the event is triggered. Rather than being triggered by a button press, however, the [on pin P0 pressed] block is triggered when you touch one of the pins—in this case, Pin 0.

Variables

The event block now needs something to do. Click on the Variables category of the Blocks toolbox, and then click the [change item by 1] block. You'll see that this block enters the Workspace with diagonal lines across it, indicating that it won't run unless it's inside a control or event block. Drag it into the [on pin P0 pressed] block, and the diagonal lines will go away (see Figure 5-9).

FIGURE 5-9: Inserting a variable block

Variable blocks are a new concept. Like the name suggests, a variable block is designed to operate on a *variable*—something which changes as the program is run. A variable can be almost anything: a number, a string of text, even the data required to draw a picture on the screen. The variable itself has two properties: its *name* and its *data*. By default, our variable is called item, which isn't very descriptive. Click on the word item within the [change item by 1] block, click on Rename variable..., and then type **touches** to make it more meaningful. This will help jog your memory when you look at the program in the future.

TIP

Variable names should always be as descriptive as possible, so their purpose within the program is obvious. There are restrictions on the names you can use, though. You can't use a name which is reserved for an instruction in the programming language you're using, a name typically can't begin with a number, and a name can't contain spaces or other symbols. To make variables with multiword names clearer to read without spaces, it's common practice to use camel case: 'Number of Fish', for example, becomes 'numberOfFish' in camel case; 'Age of User' becomes 'ageOfUser'. Camel case itself is often written as 'camelCase' to make its use more obvious.

The [change touches by 1] block does exactly what it says: each time the block runs—which will be each time the [on pin P0 pressed] event block is triggered—it takes the contents of the variable touches and increases it by one. The variable touches is *initialised* at zero; in some languages you need to do this yourself at the start of the program, but in JavaScript blocks it's done for you automatically.

The program is now, technically, complete, but not very useful. Each time you touch Pin 0, the touches variable will increase, but there's no way to see that it's happening. The program needs an output, so click on the Basic category of the Blocks toolbox and click the [show number 0] block. As before, this block will only work when it's inside a control or event block, so drag it inside the [on pin P0 pressed] block so that it is positioned below the [change touches by 1] block (see Figure 5-10).

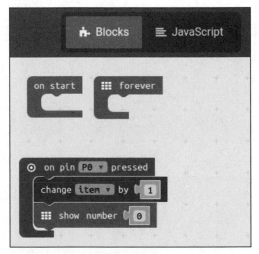

FIGURE 5-10: Inserting a show number block

You've now created your first multiline event: each time the event is triggered, the blocks inside run one at a time starting at the top and working downward until the end of the event block is reached. By chaining blocks like this, you can create programs as long and complex as you like—which is why each block has jigsaw connections on the top, for the block that comes before it, and the bottom, for the block that comes after it.

The program isn't quite finished yet, though. You've told the BBC micro:bit to increase the `touches` variable by 1—a process known in programming as *incrementing*—and display a number, but at the moment the only number it will display is 0. If you look at the [`show number 0`] block, you'll notice that the area surrounding the 0 is shaped like a jigsaw piece. This indicates that you can put a variable in there, instead of typing a set number in manually.

Click on the Variables category of the Blocks toolbox again, and then click on the [`item`] block. This block represents the `item` variable; click on the word `item` and you can change that to the `touches` variable you created earlier. Notice how the block has a jigsaw connector to the left side; this shows that it is designed to connect into another block to modify it rather than going into a control or event block as an instruction in its own right.

Drag the [`item`] block into the [`show number 0`] block over the 0. As before, the diagonal lines on the [`item`] block will disappear to show that it's now in a place where it can run. Click on the word `item` in the [`item`] block then `touches` in the drop-down list that appears to change it to a [`touches`] block, which then makes the whole block a [`show number touches`] block (see Figure 5-11).

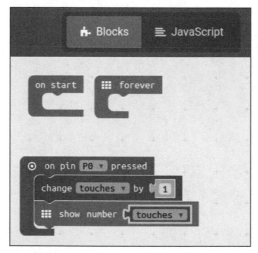

FIGURE 5-11: The finished program

The phrase `show number touches` doesn't make much sense in English, but if we break it down you can see what the computer sees: `show number` is the instruction, telling the BBC micro:bit to take a number—rather than a string—and show it on the display; `touches` isn't a number itself, but is a variable which *contains* a number. Thus, when the [`show number` touches] block runs, it will print to the display whatever number is stored in the touches variable.

If you're running the program in the simulator, click on Pin 0 now; the number 1 will appear on the BBC micro:bit's display. Click on it again, and the number will change to 2; click on it again, and it'll change to 3. The program will continue to count the number of times you've clicked on the pin until you either press the Reset button—in which case the count resets to zero again—or disconnect the power supply.

If you're running the program on a physical BBC micro:bit, touching Pin 0 isn't enough. The BBC micro:bit's ability to sense physical touch works via an electrical property known as *resistance*, which requires that the circuit—your body, in this instance—is *grounded*. Place the index finger of your right hand on the GND pin to the far right of the BBC micro:bit, and then tap on Pin 0 with the index finger of your left hand. On the first tap, the number 1 will appear; for each subsequent tap until the BBC micro:bit resets, the number will increase.

This *resistive touch sensing* is available on Pin 0, Pin 1, and Pin 2 of the BBC micro:bit. Using the inputs available from the physical buttons—Button A, Button B, and Buttons A+B— plus the three pins, it's possible to read and react to up to six inputs in total in your program: the three button inputs plus touch on Pin 0, Pin 1, and Pin 2.

When you're counting the number of apples or pears you have, you'll typically start counting at one. When programming, though, you start at zero. That's why the BBC micro:bit's three main pins are labelled Pin 0, Pin 1, and Pin 2, rather than Pin 1, Pin 2, and Pin 3. The same goes for any other numbers you're working with in your program. If you're counting the number of times a loop runs, '9' in the program means it has run ten times—counting 0, 1, 2, 3, 4, 5, 6, 7, 8, 9, for a total of ten numbers—rather than the nine you might expect. It can take a while to get used to thinking from zero, so don't worry if you forget!

TIP

There's more on using the BBC micro:bit's pins in Chapter 10, 'Building Circuits', Chapter 11, 'Extending the BBC micro:bit', and Chapter 12, 'The Wearable BBC micro:bit'.

Program 4: The Temperature Sensor

The BBC micro:bit has more inputs than simple buttons and touch-sensing pins. The most simple of these is the temperature sensor, which works exactly like a thermometer: read from the sensor and it will return a temperature, measured in degrees Celsius.

Although the temperature sensor is enough to give you a rough idea of how hot or cold your surroundings are, it's not a precision instrument. It's built into the BBC micro:bit's processor and was originally designed for monitoring the processor's temperature rather than environmental temperatures. If the BBC micro:bit were a desktop computer, that would be a problem; desktop computer processors run tens of degrees hotter than their surroundings even when idle, requiring large metal heatsinks and fans to keep them cool. The BBC micro:bit's processor, though, runs close enough to the temperature of its surroundings—known as the *ambient temperature*—that it will typically be accurate to within a degree or two.

TIP If you're looking to get the most accurate temperature reading possible, it's important to keep your program simple. If your program is working the BBC micro:bit's processor hard—doing lots of complicated sums, for instance—then the processor will begin to warm up, throwing off your readings.

Start your program in the traditional way: click on the Projects menu and then click on New so you don't lose your work on the Touch Inputs project. Click on Untitled at the bottom of the screen and then type **Temperature Sensor** to name your new project.

Click on the Input category of the Blocks toolbox, and then click on the [temperature (°C)] block to put it on the Workspace. The block will be covered in diagonal lines, indicating that it needs to be within another block to run, and is the same shape as the variable block you used to make the Touch Input program. That's because 'temperature' is a variable, one which is automatically updated every time it's used by reading the current value from the temperature sensor itself.

You can't put a variable block directly into a control block, so you'll need another block. Click on the Basic category of the Blocks toolbox and then the [show number 0] block. Drag the [show number 0] block into the [forever] block—because a thermometer that only updated once and never again wouldn't be much use—then drag the [temperature (°C)] block over the 0 in the [show number 0] block to create a [show number temperature (°C)] block (see Figure 5-12).

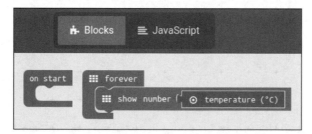

FIGURE 5-12: Reading from the temperature sensor

At this point the simulator will spring into life and start scrolling the current temperature across the BBC micro:bit's display. A slider will also appear on the simulated BBC micro:bit, set to 21°C. Because the simulator doesn't have a real temperature sensor, this slider allows you to change the reported temperature from a pretend temperature sensor. Just slide the bar up to increase the reported temperature, or slide the bar down to decrease the reported temperature.

Formatting the Output

The program is now, technically, complete, but it's not easy to read. The numbers scroll across the screen continuously with no gap, making it hard to tell the difference between 21, 12, 22, or even 212121. To fix that, click on the Basic category of the Blocks toolbox and then the [show string "Hello!"] block. Click on the word "Hello!" and then type **Celsius**—note the space at the start (see Figure 5-13).

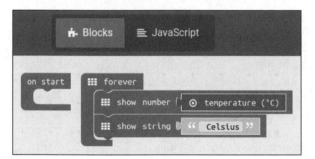

FIGURE 5-13: Formatting the program's output

The simulator will update immediately; rather than printing the temperature reading continuously with no delay or gap in the number, it will begin by printing the temperature followed by the word Celsius with a space between the two—which is the reason for including a space when you typed **Celsius** to create the [show string " Celsius"] block.

If you're running your program on a physical BBC micro:bit, touch a metal surface to discharge any static electricity you may be holding (see Chapter 2, 'Getting Started with the BBC micro:bit'). Then carefully and gently place the tip of your finger over the small black square labelled 'PROCESSOR' on the back of the BBC micro:bit. After a few seconds you should see the temperature reading start to rise. Remove your finger and after a few more seconds it will drop back down again.

Always be careful when touching components on the back of the BBC micro:bit. If you have not properly grounded yourself by touching a metal surface first, you could fire a static shock into the sensitive circuitry and potentially damage your BBC micro:bit beyond repair. **WARNING**

Program 5: The Compass Sensor

The BBC micro:bit's magnetic compass is one of the two sensors labelled on the back—the other being the accelerometer, detailed later in this chapter—and one of the most interesting. By reading the strength of the local magnetic field in three dimensions, it can work out the direction the BBC micro:bit is facing relative to magnetic north. This same technology is built into many smartphones and is used by mapping software to figure out which direction the user is facing.

The first step of this project should be obvious by now: click on the Projects menu and then click on New to create a fresh, blank project. Click on Untitled at the bottom of the screen and then type **Compass Sensor** to name your new project so you can find it again in the future. There's nothing worse than a folder full of projects all called Untitled!

Click on the Input category of the Blocks toolbox and then click on the [compass heading (°)] block. This is a variable block, the same as the earlier [temperature (°C)] block, so you're going to need another block to fit it into. Click on the Basic category of the Blocks toolbox, and then click on the [show string "Hello!"] block. Drag the [show string "Hello!"] block into the [forever] block to rid it of its diagonal lines, and ignore the simulator as it begins to repeat "Hello!" at you over and over again.

Rather than simply dragging the [compass heading (°)] block into the [show string "Hello!"] block and having it shown on the display, you're going to perform a little advanced formatting. Click on the Advanced drop-down in the Blocks toolbox, click on Text, and then click on the [join " " " "] block—it's the one with the two sets of double quotes, one on top of the other (see Figure 5-14).

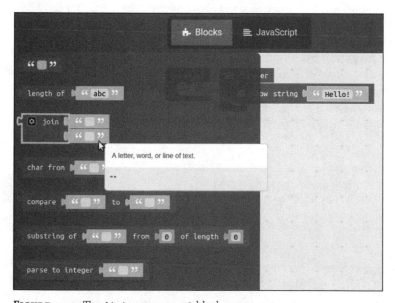

FIGURE 5-14: The [join " " " "] block

Back in the Temperature Sensor program, you used two blocks to format the output: the [show number temperature (°C)] block and the [show string " Celsius"] block, one after the other. The [join " " " "] block gives you a way of formatting your output in a similar way, but in a single block. Drag the [join " "] block to over the word "Hello!" in the [show string "Hello!"] block to create a [show string join " "] block. The simulator will stop displaying Hello!, as you just deleted that, and begin displaying the result of joining two blank strings together—which is to say, it won't display anything at all.

Drag the [compass heading (°)] block into the second, bottom, of the two double quote pairs in the [show string join " " " "] block to create a [show string join " " "compass heading (°)"] block. The simulator will begin to show the current compass heading, measured in degrees from magnetic north at 0°. To format the output, click on the first, top, of the two double quote pairs and type **Heading:** —note the space after the colon, which is there to place a gap between the string and the compass reading (see Figure 5-15).

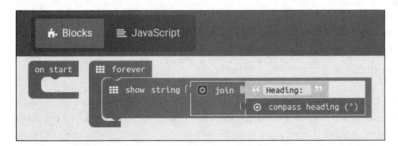

FIGURE 5-15: The completed program

If you're running the program in the simulator, you'll see that the micro:bit logo has grown a point and the label 90°. Click and hold the mouse button on the logo and drag it around to twist it; as the logo twists, the compass heading will change with 0° (magnetic north) at the top. If you're running the program on a real BBC micro:bit, simply turn the BBC micro:bit itself—taking care not to dislodge the USB or battery cable.

The accuracy of the calibration is very dependent on environmental factors. If you're calibrating the compass in an area surrounded by metal or near magnetic fields—such as on a metal desk or next to a set of speakers—the compass will not calibrate correctly. For the highest accuracy, always try to calibrate the compass in an environment as close to that in which you'll be using it as possible.

One neat feature of the `[join " "   " "]` block is that it can be nested: a `[join " "   " "]` block can have another `[join " "   " "]` block in either or both of its double quote gaps, which in turn can have another `[join " "   " "]` block or two, which in turn can have yet another `[join " "   " "]` block or two, and so on. Try it yourself by modifying the existing program to print out the word `degrees` after the compass reading without adding another `[show string "Hello!"]` block.

Program 6: The Accelerometer Sensor

The second of the two sensors labelled on the rear of the BBC micro:bit, alongside the compass, is the *accelerometer*. Whereas the compass is designed to measure magnetic fields, and thus determine which direction the BBC micro:bit is facing, the accelerometer measures relative acceleration in three planes: X, Y, and Z. As well as being able to return actual sensor values, it can be used to detect different types of motion known as *gestures*—by far the simplest way to interact with the sensor in your programs.

The accelerometer is a powerful tool. Besides being able to detect motion, it can be used to calculate the angle at which the BBC micro:bit is positioned by tracking the force of gravity pulling the BBC micro:bit toward the centre of the Earth—even when the BBC micro:bit is sat securely on a table or held in your hand. It's this feature of an accelerometer that a smartphone or tablet uses to detect when you've turned it from portrait to landscape orientation or vice versa, so that it can automatically rotate the display to match.

To begin, click on the Projects menu and then click on New to create a blank project. Click on `Untitled` at the bottom of the screen; then type **Accelerometer Sensor** to name your new project. If this part of the process seems repetitive, good. Keep repeating it until it's second nature to always create a new project and give it a name; then you won't run the risk of accidentally overwriting one of your previous programs or losing it in a sea of identically named projects.

This project doesn't use either of the `[forever]` or `[on start]` blocks. Although you've left these blocks in place previously—because if there are no blocks inside, they are simply ignored when it's time to run the program—you can tidy up the Workspace by deleting it. Right-click on the `[on start]` block and click the Delete Block option, or click and drag the block back to the Blocks toolbox, which will turn into a dustbin icon. Then do the same to the `[forever]` block.

The same right-click menu gives you some other options, too. Duplicate will create an exact copy of a block, saving you from having to keep going back into the Blocks toolbox; Add Comment will allow you to attach a note to the block to explain what it's doing for when you come back to the program later or pass it on to someone else, which is visible to the person reading the program but ignored by JavaScript Blocks itself; Help will bring up information about the block.

A well-documented program is a good program. Although you won't be instructed to type any comments in during this chapter, because they're not required to make the program run, it's a good idea to get into the habit anyway. A part of the program which makes perfect sense today might seem confusing if you come back to the program a few months down the road and could be entirely impossible for someone else to understand without guidance. Comments don't slow a program down; use them liberally, and you'll save yourself—and anyone else who uses your program—a great deal of trouble down the road.　**TIP**

Click on the Input category of the Blocks toolbox and then the [on shake] block. Like the [on button A pressed] block and the [on pin P0 pressed] block, this is an event block which stands apart from any other section of the program. This is why we could safely delete the [forever] and [on start] blocks: when the event block is triggered, the blocks within run regardless of what the main program is doing—and even when there's no main program at all.

To demonstrate that the event is triggering, it has to do something. Click on the Basic category of the Blocks toolbox and the [show icon] block. Drag it into the [on shake] block, and then click on the heart icon to bring up the icon list before finding and clicking on the surprised icon (see Figure 5-16).

The program is now ready for testing. When you added the [on shake] block to your program, the simulator added a new control: a button marked SHAKE. Click on this button to simulate shaking the BBC micro:bit, and a shocked face will appear on the display. Alternatively, move your mouse cursor over the BBC micro:bit board and wiggle your mouse to achieve the same shake effect.

The program works by using the accelerometer to look for one of a range of *gestures*—in this case, the shake gesture. If you click on the word shake in the [on shake] block, you'll see a list of the other gestures the accelerometer can track:

- **shake**—The default gesture, shake is activated by vigorously shaking the BBC micro:bit to and fro—but be careful you don't dislodge the micro-USB or battery cable or let go and send it flying across the room!

- **8g**—The 8g—and 3g and 6g—gesture triggers only when the acceleration in any given direction exceeds eight times gravity, which is pretty fast.

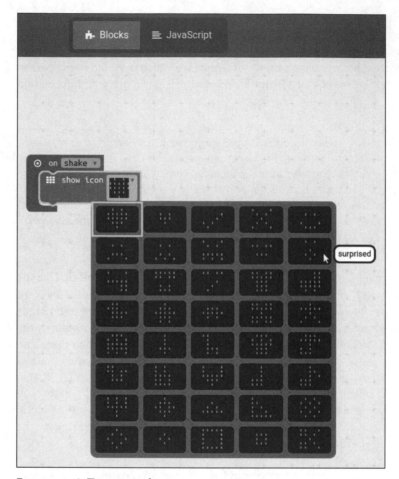

FIGURE 5-16: The surprised icon

- **logo down**—By tracking the pull of the Earth's gravity, the accelerometer can detect the orientation of the BBC micro:bit. This gesture triggers only when the BBC micro:bit is placed with the logo downward.

- **screen up**—Tracked in the same way as logo down, the screen up gesture triggers when the BBC micro:bit's display is facing upward.

- **screen down**—The opposite of screen up, this gesture triggers only when the BBC micro:bit's display is facing downward.

- **logo up**—The opposite of logo down, triggering only when the micro:bit logo is facing upward.

- **tilt right**—This gesture is based on the angle of the BBC micro:bit and will trigger when the BBC micro:bit is tilted toward the user's right.

- **free fall**—The free fall gesture triggers only when the BBC micro:bit has been dropped and is rushing toward the ground. This gesture is particularly useful when building flying robots or remote-controlled rockets.

- **3g**—The 3g gesture works in the same way as the 8g gesture but triggers at a gentler rate of acceleration.

- **6g**—The 6g gesture sits between the fast 8g and gentler 3g acceleration gestures.

- **tilt left**—The opposite of tilt right, the tilt left gesture triggers when the BBC micro:bit is tilted to the user's left.

Delays

With a few changes, the program can become a lot more interactive. Click on the Basic category of the Blocks toolbox, and then click on the [show icon] block. Drag the [show icon] block until it sits within the [on shake] block underneath the existing [show icon] block; then click on the heart icon and find asleep in the list.

If you press the SHAKE button on the simulator now, you'll briefly see the surprised icon before the asleep icon takes over. If you were to continuously press the SHAKE button—the simulator equivalent of continuously shaking the BBC micro:bit without stopping—the 'face' would change from the surprised icon to the asleep icon and back again over and over relatively quickly. That's not very realistic; it takes time to get back to sleep when you've been shaken awake!

To fix that, you're going to need to add a *delay* between the two [show icon] blocks. A delay does exactly what its name suggests: it pauses the program for a set length of time, delaying the execution of the next block. Click on the Basic category of the Blocks toolbox, and then click on the [pause (ms) 100] block. Drag it into the [on shake] block so it sits between the two [show icon] blocks; the [show icon] blocks will automatically move apart to make room (see Figure 5-17). Click on the 100 number in the [pause (ms) 100] block and change it to read 1000 to create a [pause (ms) 1000] block.

This time, if you click the SHAKE button—or, if you're running the program on a physical BBC micro:bit, shake the BBC micro:bit—you'll see the surprised icon for longer before the asleep icon takes over. Increase the number in the [pause (ms) 1000] block and the surprised face will stay on the display for longer; decrease it and the asleep face will appear sooner.

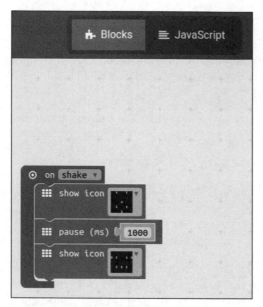

FIGURE 5-17: The completed program

The number in the [pause (ms) 1000] block is a measure of how long the program should delay in *milliseconds*, or thousandths of a second. A value of 1000, then, is equivalent to one second; 2000 would be two seconds, and 500 would be half a second. Because processors like the one powering the BBC micro:bit operate so much faster than humans, performing millions of operations every second, programs frequently need delays in them to allow the human operator to catch up.

As an experiment, see if you can extend the program still further by adding an [on start] block back in and having the BBC micro:bit start on the asleep icon, rather than starting entirely blank until the first shake.

Reading Raw Accelerometer Data

The gesture system is by far the easiest way of working with the BBC micro:bit's accelerometer, but it's not the only way. You can also read the raw data into your program as numbers to handle yourself. Create a fresh project by clicking on the Projects menu, and then click on New. Click on Untitled at the bottom of the screen, and then type **Accelerometer Sensor Data** to name your new project and keep it separate from the earlier 'Accelerometer Sensor' gesture-triggered project.

Start by deleting the [on start] block by either right-clicking and clicking Delete Block or dragging it to the Blocks toolbox. Next, click on the Basic category of the Blocks toolbox and then click on the [show string "Hello!"] block. Drag the [show string "Hello!"] block into the [forever] block, and ignore the simulator because it again greets you with an infinitely scrolling message.

Click on the Advanced drop-down option in the Blocks toolbox if the Advanced categories aren't already visible, click on the Text category, and then click on the [join " " " "] block. Drag the [join " " " "] block into the [show string "Hello!"] block over the word "Hello!" Click in the first, top, double quote section of the [join " " " "] block and type **X:** to create a [join "X:" " "] block.

Click on the Input category of the Blocks toolbox and then click on the [acceleration (mg) x] block. Drag this into the second, bottom, double quote section of the [join "X:" " "] block to create a [join "X:" "acceleration (mg) x"] block. The simulator will now start scrolling the string "X:" followed by the current measurement from the accelerometer across the X axis. Try moving the simulated BBC micro:bit with your mouse cursor, and watch the number change.

The X axis is only one of three axes that make up the accelerometer's three-dimensional measurements, though. To get at the others, right-click on the show string section of the now quite large [show string join "X:" "acceleration (mg) x"] block and click Duplicate from the menu that appears. Drag the new block into the [forever] block underneath the existing block, and then right-click the show string section of the new block and choose Duplicate again before dragging the third block underneath the existing two.

Your program is now displaying the accelerometer data three times, but it's the same X axis data each time. Click on the "X:" part of the second full block, in the top section of the [join "X:" "acceleration (mg) x"] block, and change it to "Y:". Next, click on the "x" of "acceleration (mg) x" and change it to "y". The whole second block should now read [show string join "Y:" "acceleration (mg) y"]. Do the same for the third block, but change the "X:" and "x" to "Z:" and "z" instead so it reads [show string join "Z:" "acceleration (mg) z"] (see Figure 5-18).

The BBC micro:bit will now read out all three of the axes, one after another (the X axis, followed by the Y axis, and then the Z axis), before returning the X axis again. Try moving the simulator to see the numbers change, or flash the program onto a real BBC micro:bit and see how the angle and position of the BBC micro:bit are reflected in the numbers—measured in *microgravities*—shown on the display. With the BBC micro:bit facing you, tilting it left will reduce the number reported on the X axis, while tilting it right will increase it. Tilting it so the top is closer to you will reduce the number reported on the Y axis, while tilting it so the top is further away will increase it. Finally, lifting the BBC micro:bit up will increase the number reported on the Z axis, while lowering it down will reduce it (see Figure 5-19).

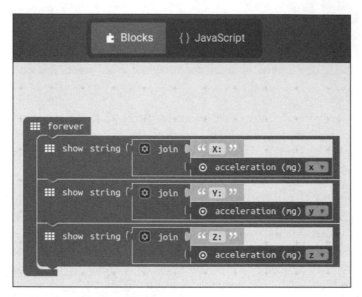

FIGURE 5-18: The completed program

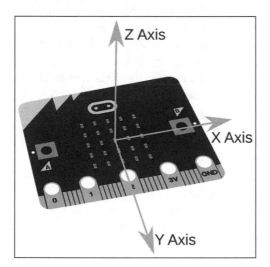

FIGURE 5-19: The accelerometer's three axes

Program 7: The Fruit Catcher Game

The programs in this chapter have been relatively simple, doing only one thing at a time, as a means of introducing some of the base concepts you'll need to know when programming the BBC micro:bit in JavaScript Blocks. This final program, though, is considerably more complex: a game, which has the player—in the form of a single glowing pixel at the bottom of the display—attempt to catch ever-faster falling fruit—another pixel, starting at the top of the display and falling downward—before it hits the bottom of the display.

This program uses a category of the Blocks toolbox's Advanced section not previously seen: the Game category. This is a category of blocks designed specifically for creating simple games, with built-in tools for keeping track of the score, displaying a 'game over' screen, creating single-pixel *sprites*—objects which can be moved around the display—and detecting when a sprite has collided with another sprite or the edge of the display.

The Fruit Catcher game is relatively simple but still quite a hefty program. If you've been working through this chapter nonstop, now might be a good time to take a break away from the screen so you can come back refreshed.

As ever, the game starts the same way as any program: creating a new project by clicking on the Projects menu and then clicking New and giving it a name by clicking on Untitled and typing **Fruit Catcher**. You'll need both the [forever] block and the [on start] block, so don't delete these. If you already deleted them before reading all of this paragraph, just click the Undo button to bring them back or create new ones from the Basic category in the Blocks toolbox.

If you'd prefer to type the program out in one go and then read about what each block does afterward, you'll find a full copy of the program's blocks in Appendix A, 'JavaScript Blocks Recipes'. **TIP**

The Setup

The game should start by setting up the player character sprite and initialising the score. Click on the Variables category of the Blocks toolbox, click the [set item to 0] block, and then drag it into the [on start] block. Click on the word item and then click Rename Variable to give it a more descriptive name: player.

To make the player variable a sprite, click on the Advanced drop-down of the Blocks toolbox if the Advanced categories aren't already visible, click the Game category, and then click the [create sprite at x: 2 y: 2] block. Drag this block and attach it to the [set player to 0] block over the 0 to create a [set player to create sprite at x: 0 y: 0] block. Finally, make sure the first number, next to X, is 2, and change the second number, next to y:, from 0 to 4.

The numbers in the [create sprite at x: 0 y: 0] block are important; they control where the sprite appears on the BBC micro:bit's display. Every pixel on the display has a location on the horizontal X axis and the vertical Y axis. Figure 5-20 demonstrates this, giving the X and Y coordinates as 'X,Y' for each of the 25 on-screen pixels.

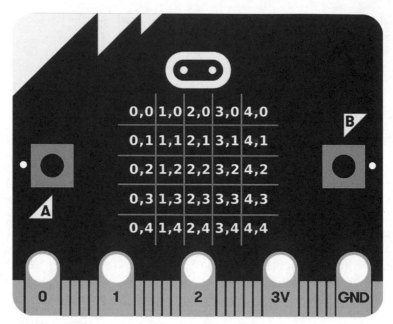

FIGURE 5-20: The BBC micro:bit Display Coordinates

By telling the [create sprite at x: 2 y: 4] block to create the player sprite at position 2 on the X axis and 4 on the Y axis—coordinate 2,4 in Figure 5-20—the sprite is created in the middle column of the bottom row of the display.

Next, click on the Game category of the Blocks toolbox, click the [set score 0] block, and then drag it into the [on start] block underneath the [set player to create sprite at x: 2 y: 4] block. This will initialise the game's score at zero. You could, if you wanted, cheat at this point and have the game start on a score of 10, or 100, or any other number you like, but where's the fun in that?

Finally, click on the Variables category of the Blocks toolbox and click on the [set item to 0] block, and drag it into the [on start] block under the [set score 0] block. Right-click the [set item to 0] block, click Rename Variable, type **delay**, and then click on the 0 and change it to 1000 (see Figure 5-21). This variable will control how fast the game is when it starts. If you'd like a challenge, you can make the variable lower; to make the game easier, make it higher.

The Main Program Loop

The next stage of the process is to make the game itself. Click on the Variables category of the Blocks toolbox, then the [set item to 0] block, and then drag it into the [forever] block. Right-click the [set item to 0] block, click Rename Variable, and then type **fruit**. As with the player sprite, click on the Game category of the Blocks toolbox, the [create sprite at x: 0 y: 0] block, and then drag it over the "0" of the [set fruit to 0] block to create a [set fruit to create sprite at x: 2 y: 2] block.

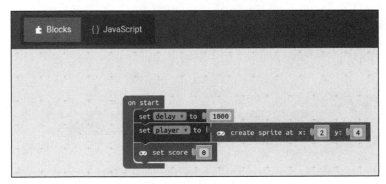

FIGURE 5-21: The finished [on start] block

If the fruit appeared in the same place every time you played the game, it wouldn't be much of a game. To make things more interesting, click on the Math category of the Blocks toolbox, click on the [pick random 0 to 4] block and drag it over the first 2 of the [set fruit to create sprite at x: 2 y: 2] block next to x, and then change the second 2, next to y:, to 0.

The [pick random 0 to 4] block does exactly what it says: picks a random number between zero and the second number in the block, which defaults to four. With a range of five numbers, the default [pick random 0 to 4] block maps perfectly to the X or Y axis of the display. That's how it's used here: when the game creates the fruit sprite, it will position it at the top (Y: 0) of the display but choose a random location in that top row (X: 0 to 4, depending on what the random number generator picks).

At the rate the BBC micro:bit processor runs through blocks, a game where the fruit falls as quickly as possible would be many millions of times faster than a human could play; you'd see nothing but the 'game over' screen. To fix that, click on the Basic category of the Blocks toolbox, click on the [pause (ms) 100] block, and drag it into the [forever] block under the existing block. Rather than setting a manual delay, though, use a variable so the game can get faster as it progresses: click the Variables category of the Blocks toolbox, click the [item] block, drag it over the 100 to create a [pause (ms) item] block, and then click the word item and click delay from the list that appears to use the delay variable you created in the [on start] block.

At this point, the program should look like Figure 5-22.

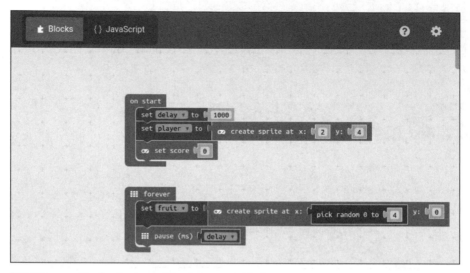

FIGURE 5-22: The beginnings of the [forever] block

Conditional Loops

So far, you've only worked with a single type of loop: the infinite loop, which runs forever. A more powerful type of loop is the *conditional loop*, which tests for a particular condition—such as whether a variable is equal to another variable—and runs only when the condition is true.

Click on the Loops category of the Blocks toolbox, click on the [while true do] block, and then drag it into the [forever] block under the existing blocks. By default, the [while true do] block is an infinite loop just like the [forever] block itself: the condition it is testing is whether true is 'true'—and because true is, indeed, true, the condition itself always returns true and the loop never ends.

To change that, click on the Logic category of the Blocks toolbox, click on the [0 = 0] block, and then drag it over the word true in the [while true do] block to create a [while 0 = 0 do] block. Next, we want to make it so the loop only runs when the fruit isn't already on the bottom row of the display. Click the Game category of the Blocks toolbox, click on the [item x] block, drag it over the first 0 of the [while 0 = 0 do] block, click the word item and choose the fruit variable, and then click the x and change it to y so you're looking at the fruit sprite's position on the Y (vertical) axis of the display rather than the X (horizontal) axis.

To check whether the fruit has finished falling, click on the last 0 of what is now a [while fruit y = 0 do] block, type **4** to represent the fifth (counting from zero) row of the display, and then finally click on the = symbol and change it to a < symbol to create the finished [while fruit y < 4 do] block (see Figure 5-23). The block now gets the vertical position of the fruit and checks to see if it's less than four; if it is, the fruit hasn't reached the bottom of the screen and the blocks inside the loop run; if it is equal to or greater than four, the fruit has reached the bottom of the screen and the blocks inside the loop are skipped.

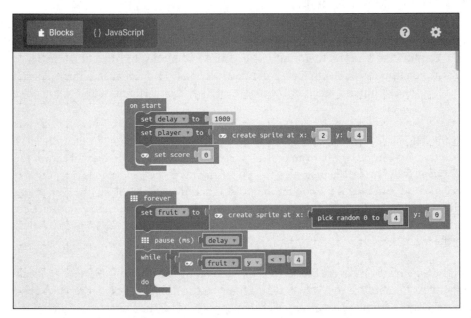

FIGURE 5-23: The conditional loop

At the moment, though, the loop doesn't actually do anything. Click on the Game category of the Blocks toolbox, click on the [item change x by 1] block, and then drag it inside the [while fruit y < 4 do] block next to the word do. Click on item and change the variable to fruit, and then click on x and change it to y. Now, every time the [fruit change y by 1] block runs—which is whenever the loop allows it to, in other words, only when the fruit's position on the Y axis of the display is less than four—the fruit's Y axis position will be increased, or incremented, by one. The result: the fruit will seem to be 'falling' down the display.

The falling process needs to be slowed down to make the game fun, though, so right-click on the [pause (ms) delay] block you created earlier and then click Duplicate. Drag the duplicated block down into the [while fruit y < 4 do] block underneath the [fruit change y by 1] block. Now the game will wait for the number of milliseconds specified in the delay variable—which you initialised at 1000 for a delay of one second—before the fruit drops down to the next column.

Conditional Statements

Loops aren't the only blocks to which conditions can be attached. A powerful programming tool is the 'if, then, else' *conditional control flow*. Using an 'if, then, else' control, you can test to see if a condition is true and then run a particular set of blocks, or run a different set of blocks if the condition is false. Unlike a conditional loop, a conditional statement doesn't loop: whatever blocks run as a result of the condition, they'll only run once unless you place them inside a loop block.

Click on the Logic category of the Blocks toolbox, click on the [if true   then   else] block, and then drag it into the bottom of the [forever] block. This conditional allows you to create two lists of blocks: the first, next to true, run only if the condition is true; the second, next to else, run only if the condition is false. There's a simpler block, the [if true then] block, which you'll use later in the program, that does away with this second section. If the condition is false in an [if true   then] block, then it simply moves on to the next block in the list.

Click on the Game category of the Blocks toolbox, click on the [item touching   ?] block, and drag it next to the if of the [if true   then   else] block. Click on the word item and choose player from the list of variables that appears. Click the Variable category of the Blocks toolbox, click the [item] block, drag it into the gap between the word touching and the ? symbol of the [player touching   ?] block, and then click the word item to change it to the fruit variable. The final block should read [player touching fruit?].

This [player touching fruit?] block is the test for the conditional; it checks whether the player sprite, created at the start of the game, is touching the fruit sprite when the fruit sprite has reached the bottom of the screen. If the player sprite is touching the fruit sprite, the player has 'caught' the fruit; if not, the player has 'missed' the fruit.

If the player has caught the fruit, the player's score needs to be increased. Click the Game category of the Blocks toolbox, click the [change score by 1] block, and drag it into the then portion of the [if player touching fruit   then   else] block.

If the player has missed the fruit, the game is over. Click the Game category of the Blocks toolbox, click the [change score by 1] block, and drag it into the else portion of the [if player touching fruit then else] block. Read as a whole, this collection of blocks should now read [if player touching fruit then change score by 1 else game over].

Finally, you need two more blocks to finish off the main portion of the game. Click the Game category of the Blocks toolbox, click the [item set x to 0] block, and then drag it into the [forever] block under all the other blocks you've created. Click on the word item and change it to the fruit variable, and then click on x and change it to the brightness property. Now when the player has caught the fruit sprite, its brightness will be set to zero, hiding it from view.

Place the last block of the main program by clicking on the Variable category of the Blocks toolbox, then the [set item to 0] block. Drag the block into the [forever] block underneath all the other blocks. Click on the word item and change it to the delay variable. After that, click on the Math category of the Blocks toolbox, click on the [0 - 0] block, and then drag it over the 0 in the [set item to 0] block. Click on the Variable category of the Blocks toolbox, click on the [item] block, drag the block over the first 0 of the [0 - 0] block you just created, and then click on the word item and change it to delay to create a [delay - 0] block.

Right-click on the [delay - 0] block just under, but not on, the word delay, and then click Duplicate. Drag the duplicated block over the second 0 to create a [delay - delay - 0] block, and then click on the - symbol after the second delay and change it to a ÷ symbol. Last but not least, click on the last 0 and change it to 10.

The finished block should read [set delay to delay - delay ÷ 10], and it's this block that is responsible for speeding the game up after every successful catch of the fruit. The first part of the block tells JavaScript Blocks to change the value of the delay variable to the outcome of a short equation; the second tells it to take something away from the current value of the delay variable; the last part tells it to take away the current value of the delay variable divided by 10. The result: the delay variable shrinks by 10 percent each time the fruit is caught, making the game 10 percent faster each round.

You've now finished the main program loop (see Figure 5-24). Although the game may appear to be running in the simulator, there's a problem: you can't yet control the player sprite.

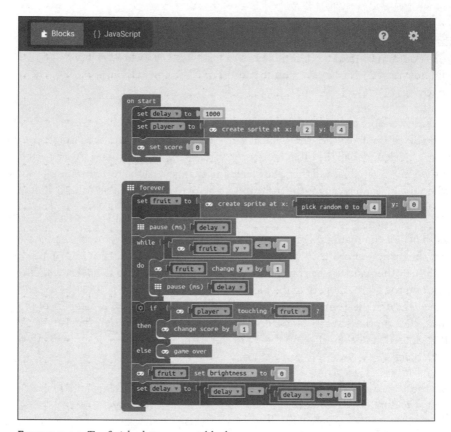

FIGURE 5-24: The finished [forever] block

The Control Events

To catch the fruit, the player needs to be able to move left and right along the bottom of the BBC micro:bit's display. Handily, there's Button A to the left of the display and Button B to the right—just what you need for controlling the player sprite. Making these control events is effectively a reprise of the Button Inputs program you wrote earlier in the chapter, but with a little extra logic to help things along.

First, click on the Input category of the Blocks toolbox, and then click on the [on button A pressed] block to create an event block before dragging it to a clear area of your Workspace. Next, click on the Logic category of the Blocks toolbox, click on the [if then] block, and drag it into the [on button A pressed] block.

To create the conditional statement for the [if then] block, click the Logic category of the Blocks toolbox, click the [0 < 0] block, and then drag it next to the word if to connect

it to the [if   then] block. Click on the Game category of the Blocks toolbox, click the
[item x] block, and then drag it over the first 0 of the [0 < 0] block before clicking the
word item and changing it to the variable player. Finally, click the < (less than) symbol and
change it into a ≥ (greater than or equal to) symbol.

The [if player x ≥ 0   then] conditional exists to stop the player sprite leaving the left
edge of the display. It checks the player's current position on the X (horizontal) axis. If it's
greater than or equal to 0, representing the left edge of the display, the player should be
allowed to move farther left; if not, the player is already at the leftmost edge and should not
be allowed to move farther left.

To actually make the player sprite move, click the Game category of the Blocks toolbox, click
the [item change x by 1] block, and then drag it into the [if player x > 0
then] block next to the word then. Click the word item, change it to the variable player,
and then click on the number 1 and replace it with –1 to create a [player change x by
-1] block. Now when the player presses Button A but isn't already at the left edge of the
screen, the player's position on the X axis is decreased by one—or *decremented*, the opposite
of incremented—moving the sprite one pixel to the left.

To save time in making the same event block for moving right, right-click on the [on button
A pressed] block and click Duplicate. Drag your duplicated collection of blocks to a clear
area of the Workspace, and then make the following changes: click the A of the [on button
A pressed] block and change it to B; click the > symbol of the [if player x ≥ 0
then] block to < and the 0 to 4; then change the –1 of the [player change x by -1]
block to 1.

Now the game has full control: the first event block watches for Button A being pressed,
checks that the player has room to move to the left, and then moves the player sprite leftward one column. The second event block watches for Button B being pressed, checks that
the player has room to move to the right, and then moves the player sprite rightward one
column. The game is complete and should look like Figure 5-25.

Before trying out your new game, check that the program has been entered correctly by comparing it to the version found in Appendix A and making any changes necessary. Once you're
sure the program is correct, you can play the game on the simulator. Just click the Reset button, shaped like a pair of arrows chasing each other in a circle, and the game will start. Move
the player sprite, at the bottom of the screen, with Button A and Button B, and try to catch
the fruit sprite as it falls from the top to the bottom. Miss the fruit and it's game over, and
you'll see a flashing animation, a 'GAME OVER' message, and your score. To restart the
game, just click the Reset button again—and that will work on both the simulator and a real
BBC micro:bit.

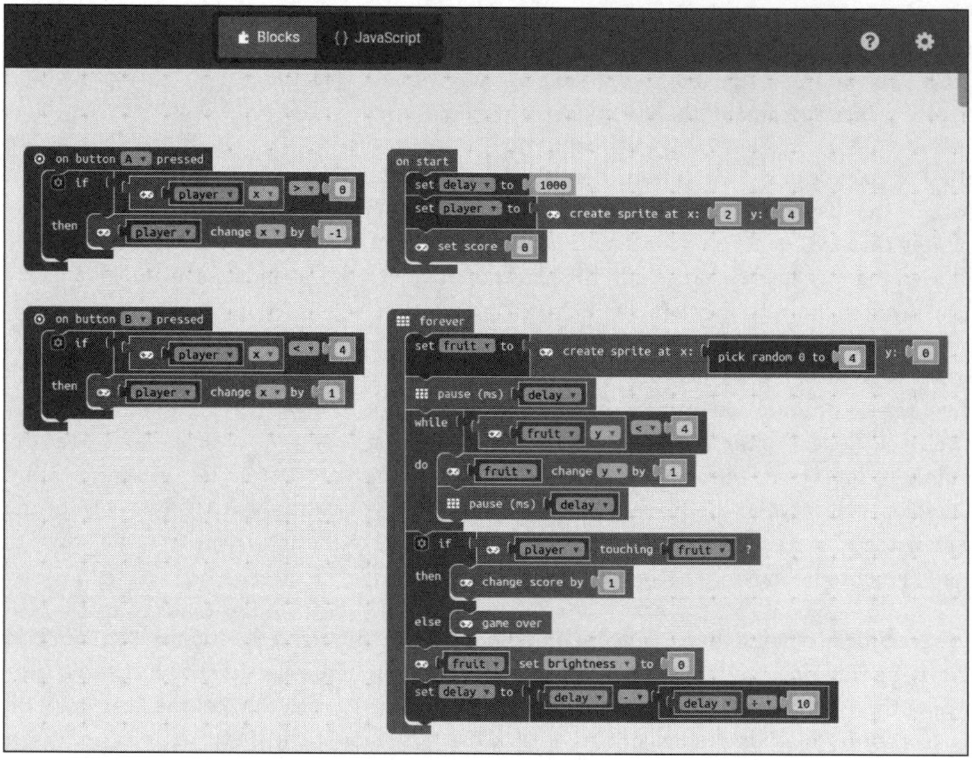

FIGURE 5-25: The finished game

To improve the game—and your programming skills—try making some modifications. Instead of controlling the player sprite with the buttons, how about using the accelerometer to watch for tilting and give the game motion control? Try adding multiple fruit which all fall down the screen at once to increase the challenge.

Further Steps

Now that you've finished the chapter on programming in JavaScript Blocks, why not try the same exercises in JavaScript or Python in the following chapters? If you'd prefer to stick with JavaScript Blocks, click on the Projects menu of the JavaScript Blocks Editor followed by the Make tab to bring up a range of additional projects to try, from a touch-sensitive 'love meter' to a robot built from a milk carton.

Chapter 6
JavaScript

In this chapter

- An introduction to programming in the JavaScript Editor
- 'Hello, World!': Your first JavaScript program
- Programs for reading from button inputs, touch inputs, and the temperature, accelerometer, and compass sensors
- Fruit Catcher: A simple game to program and play on your BBC micro:bit

THE JAVASCRIPT LANGUAGE is a popular choice for beginners, with its easy-to-understand format and layout. It's also one of the most common languages underpinning the World Wide Web. When you visit a web page and it has interactive content, that content is almost always powered by JavaScript. If you've previously worked through Chapter 5, 'JavaScript Blocks', then you've already had an introduction to how the language works: Programs created in the JavaScript Blocks Editor are invisibly transformed into JavaScript code as you build them, and it's this code that is compiled and run on your BBC micro:bit or in the simulator. Technically speaking, the language used on the BBC micro:bit is a version of TypeScript on top of JavaScript, but this combination is most commonly referred to as simply 'JavaScript'.

As in Chapter 5, in this chapter you are introduced to the JavaScript Editor, powered by Microsoft MakeCode, and learn to write a series of programs that make the most of the BBC micro:bit's capabilities. Finally, you're able to make a quick-fire game in which you're given the task of catching ever-faster falling fruits before they hit the ground. These programs are the same between Chapter 5, 'JavaScript Blocks', Chapter 6, 'JavaScript', and Chapter 7, 'Python', allowing you to easily compare and contrast the three languages to find the one that suits you best.

Introducing the JavaScript Editor

The JavaScript Editor is an all-in-one development environment created specifically for the JavaScript language. Running entirely within the web browser of any modern computer, the JavaScript Editor doesn't require you to install software on your computer. All you need to do to load the JavaScript Editor is open the web browser of an Internet-connected computer, type **makecode.microbit.org** into the address bar and press the Enter key, and then click on the Editor toggle in the centre of the top menu bar to switch from JavaScript Blocks Editor mode to JavaScript Editor mode (see Figure 6-1). Once the editor has loaded, it runs directly on your computer even if you disconnect from the Internet.

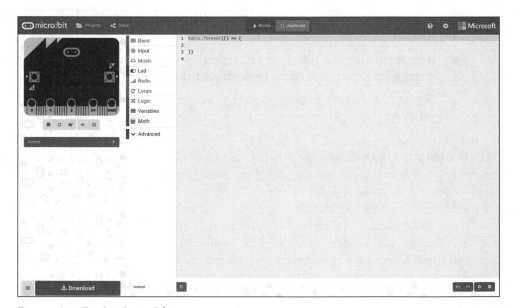

FIGURE 6-1: The JavaScript Editor

The main features of the JavaScript Editor follow:

■ **The Projects menu**—Located at the top-left and looking like a folder icon, use this menu to create a new, blank project, or to load a previous project. These project files are stored on your computer, in your web browser's local storage area, so are available only to you unless you choose to share them. Be aware that clearing stored data using your web browser's 'clear history' option will delete these files, so make sure to save any you wish to keep using the Save icon. The Projects menu also offers access to sample projects under the Make tab, and code samples under the Code tab.

- **The Share button**—Located next to the Projects menu and shaped like a circle with two further circles attached on sticks, use this button to publish and share a program you've created in the JavaScript Editor.

- **The Editor toggle**—Looking like a jigsaw piece next to some lines of writing and located in the middle of the top menu bar, use this to toggle between JavaScript and JavaScript Blocks modes. The former mode will be used throughout this chapter; the latter mode is explained in full in Chapter 5.

- **The Help button**—Found to the right of the top menu bar and shaped like a question mark in a circle, use this button to access the JavaScript Editor's built-in help, including a getting started tutorial—also available using the orange 'Getting Started' button to the right—and sample projects.

- **The Settings button**—Located next to the Help button, the Settings button is shaped like a cog. Use this button to change project settings, add additional packages such as libraries for using additional hardware with your BBC micro:bit, delete a project, or delete all your saved projects. You can also view the Editor's privacy policy, terms of use, and provide feedback from here.

- **The simulator**—Located to the left on a high-resolution screen or to the bottom left on a lower-resolution screen, the simulator is more than just a picture of a BBC micro:bit. When you write your program, it automatically begins running in the simulator exactly like it would on the BBC micro:bit itself. This allows you to test your program without having to download the compiled hex file and flash it to your BBC micro:bit. The simulator is fully interactive: to simulate pressing Button A, just click the picture of Button A. If your program uses one of the sensors, the simulator adjusts to allow you to use those too, adding a temperature slider for the thermometer or letting you change its angle for the accelerometer or compass.

- **The Blocks toolbox**—Located in a strip at the left of the screen, the Blocks toolbox can be used to remind you of how JavaScript instructions are written and reduce the amount of typing you need to do as you build your programs. Blocks of code are dragged out of the Blocks toolbox into the program listing, where they are instantly converted into one or more lines of JavaScript code. Each block is separated into a series of categories: Basic, Input, Music, LED, Radio, Loops, Logic, Variables, and Math, with additional blocks available by clicking the Advanced drop-down menu.

- **The program listing**—Taking up the bulk of the screen and with line numbers to the left, the program listing area contains the JavaScript program on which you are currently working. When you start a new project, you'll already have four lines: one reading `basic.forever(() => {`, one reading `})`, and two which are blank.

- **The Download button**—When you've finished your program and want to run it on your BBC micro:bit, just click the Download button at the bottom of the screen and the JavaScript Editor compiles it and downloads the hex file for you to drag and drop onto the MICROBIT drive (see Chapter 3, 'Programming the BBC micro:bit').

- **The project name**—The first thing to do when you start a new project is to give it a descriptive name. Always start by renaming a project from the default Untitled to make it easier to find in the future.

- **The Save icon**—Located next to Project Name, the Save icon allows you to save a copy of your program to your computer. If you want to be able to take a copy home from school or vice versa, the Save icon lets you do this. Saved projects can be loaded into the JavaScript Editor by clicking on the Projects menu and then Import File.

- **The Undo and Redo buttons**—Shaped like curled arrows, the Undo and Redo buttons allow you to undo a mistake by clicking on the leftmost arrow and then redoing it by clicking the rightmost arrow—just in case undoing the mistake was the real mistake!

- **The Zoom buttons**—Found at the bottom right of the screen, use these buttons to zoom in and out of the workspace to make the program listing text larger or smaller. For more complex projects, zooming out allows you to see more of the project onscreen at once; if you find the program listing hard to read, zooming in can make things clearer.

Program 1: 'Hello, World!'

'Hello, World!' is the traditional first program for any language. It's basic by design but offers a first glimpse of what a given language will be like to program in with the bonus of a quick payoff in the form of a message proving that your program is alive and well.

Start by opening your browser and going to makecode.microbit.org to load the JavaScript Blocks Editor; then click the Editor Toggle to switch to JavaScript Editor mode. If you've been working on projects previously, including projects in JavaScript Blocks, click the Projects menu and New to load a new, blank project. Click on the box labelled Untitled at the bottom of the screen to rename your program and type **Hello World** to give it its new name.

'Hello, World!' is a simple program that requires just one line of code in addition to the two lines you have on your screen. Click on Line 4 of the program listing, which is labelled with a

4 on the left side, or move the cursor with the arrow keys on your keyboard so that Line 4 is highlighted. Then type the following:

```
basic.
```

As soon as you type the period, a box appears containing all the *functions* available under the category Basic. To save time on typing, you could find the function you need in the list and click it to make it appear in your program listing. For now, though, continue typing until Line 4 reads like this:

```
basic.showString("Hello, World!")
```

As another way of reducing the amount of typing you have to do to complete your program, you can click on the Basic category of the Blocks toolbox, click on the [basic. showString("Hello, World!")] block, and then use the keyboard to delete "Hello!" and replace it with "Hello, World!" (see Figure 6-2).

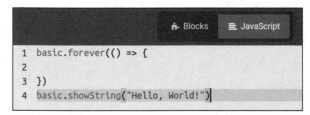

```
1 basic.forever(() => {
2
3 })
4 basic.showString("Hello, World!")
```

FIGURE 6-2: The basic.showString instruction in the program listing

---

If you make a mistake at any point while writing your program, simply click the Undo button at the bottom right of the screen to send it back to the Blocks toolbox. Alternatively, right-click on the block and click the Delete Block option. **TIP**

In JavaScript, everything is *case sensitive,* meaning it always has to be typed in the exact right mix of uppercase and lowercase. The instruction basic.showString is valid, but Basic. showString, BASIC.SHOWSTRING, basic.showstring, and any other variant simply does not work. **TIP**

You may have noticed that the line of code you've just written uses different colours. This is known in programming as *syntax highlighting* and helps you see any mistakes you may have made at a glance. The instruction itself, the basic.ShowString portion of the code, is displayed in blue; the message that the instruction is using, "Hello, World!", is displayed in red; the brackets, meanwhile, are displayed in black.

Try deleting the last double-quote from Line 4, so it reads like this:

```
basic.showString("Hello, World!)
```

Notice how the colours have changed: the last bracket, which was black before, is now red, indicating that the JavaScript Editor thinks it's part of the string. Without the last double quote to tell the JavaScript Editor where the string ends, though, the program no longer works; you've introduced a *bug* into the program. Handily, the JavaScript Editor has another trick up its sleeve to help you track down bugs: a squiggly red line underneath any line of code the Editor can't understand, which now includes the broken Line 4. If you see this, it means there's something wrong with the line of code—or, possibly, a line directly above it. Check carefully to see where the mistake might be, and hover your mouse cursor over the line without clicking to receive a helpful error message from the JavaScript Editor itself (see Figure 6-3).

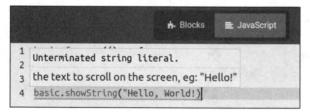

FIGURE 6-3: The JavaScript Editor highlighting a code bug

TIP　The JavaScript Editor's pop-up tips aren't limited to lines containing bugs: Hover your mouse cursor over any line containing an instruction to see a brief description of the instruction and how to use it. Try it by hovering your mouse cursor over Line 1 in your program.

Put the double-quote back into Line 4 so it reads like this:

```
basic.showString("Hello, World!")
```

The squiggly underline indicating a bug will disappear, and at this point the simulator fires into life. By typing in that one line of code and fixing a bug, you've completed your first program. The message "Hello, World!" will scroll across the simulator's display, just like it would on a real BBC micro:bit (see Figure 6-4). Feel free to experiment at this point: change Line 4 to scroll a message of your own, such as your name or your favourite colour.

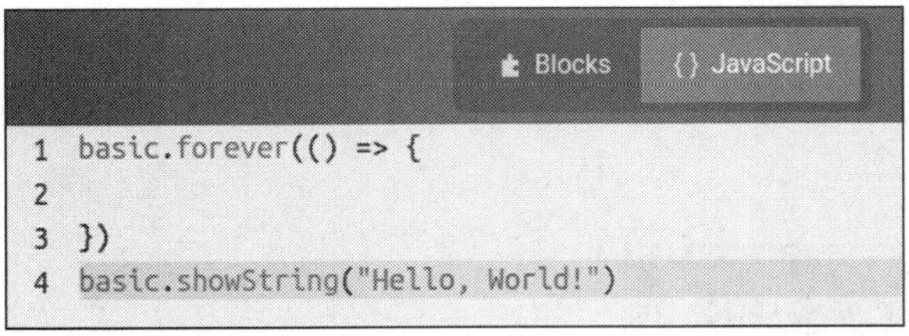

```
   ♣ Blocks    {} JavaScript
1  basic.forever(() => {
2
3  })
4  basic.showString("Hello, World!")
```

FIGURE 6-4: The simulator displaying part of the scrolling message

> In programming, messages like this are called *strings* and can be made up of letters, numbers, and some symbols. Although a string is great for displaying messages on a screen, you're limited in how you can work with it; the string "1", for example, can't be used in a mathematical equation because although it looks like a number to us: it's simply a string to the programming language. For mathematical equations, there are other types: *integers* are whole numbers and *Booleans* are logic elements which can be True or False—equivalent to 1 and 0, respectively.
>
> **TIP**

The simulator runs in the background automatically, and every time you change a line of code or add a new line of code to your program, it restarts the program to show you what your change did. Sometimes when you're writing a longer program, this can get distracting. To turn the simulator off, just click the left Stop button, marked by a square stop symbol, underneath it; to turn it back on again, click the Play button, marked by a triangular play symbol, that has appeared in its place.

Whatever message you decide to scroll, you'll notice it only scrolls once and then the simulator stops. This is the result of placing your line of code at the bottom of the program on Line 4, outside the brackets opened on Line 1 and closed on Line 3. Whatever is outside these brackets only runs once, when the program begins. You can make the message appear again if you reset the simulator by pressing the icon shaped like two bent arrows underneath the picture of the BBC micro:bit, but if you don't want to have to keep doing that, you're going to need a method of telling the program to repeat.

Loops

Move the code you've written by clicking on the number 4 at the start of Line 4 and pressing Ctrl+X (Cmd+X on macOS) on the keyboard to copy it before clicking anywhere in the blank line next to the 2 at the start of Line 2 and pressing Ctrl+V (Cmd+V on macOS). You're left with a blank Line 4 and your code in Line 2 (see Figure 6-5). The JavaScript Editor automatically moves it inward by four spaces; this is known as *indentation*, and it is a way to easily see how lines of code in a program *nest* within each other. Indentation is handled automatically by the JavaScript Editor; although it's not necessary for the program to run—the code could be entirely flat against the left edge of the program listing and still work perfectly—it makes it a lot more readable and bugs easier to track down. When you've moved your line of code, watch the simulator. The program restarts, as it always does when you make a change, but rather than scrolling the message once and then stopping the message, it repeatedly scrolls forever.

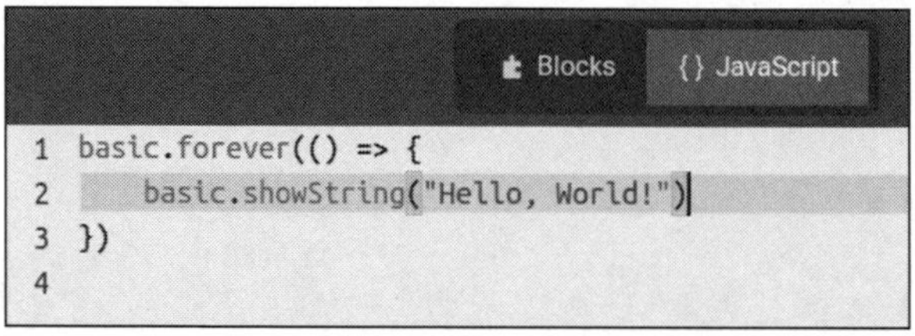

```
                                    Blocks    {} JavaScript
1  basic.forever(() => {
2      basic.showString("Hello, World!")
3  })
4
```

FIGURE 6-5: Looping the basic.showString message

In programming, this is known as a *loop*. When the program reaches the end of its instructions—in this case, when it has finished scrolling the message from the basic.showString("Hello, World!") instruction you moved to Line 2—it returns to the beginning of the loop and starts all over again. As the name implies, the basic.forever loop loops forever; in programming, this is known as an *infinite loop*.

If you want the program to stop, you can either remove the basic.showString("Hello, World!") instruction on Line 2 from the basic.forever loop opened on Line 1 and closed on Line 3 or click the square Stop button under the simulator.

> **TIP** All loops need to be opened and closed, just like all strings need to be between paired double quotes. The { symbol at the end of Line 1 indicates the start of the loop, and any line that follows this becomes a part of the loop and runs forever. The } symbol at the start of Line 3 marks the end of the loop, and any line after this only runs once and does not form part of the loop.

If you want to see how the program runs on a real BBC micro:bit, click the Download button to compile it into a hex file and drag it to the micro:bit's drive, as described in Chapter 3.

Program 2: Button Inputs

'Hello, World!' demonstrated how to get an output from the BBC micro:bit via the screen, but there's another key feature of most programs: inputs. Thankfully, the BBC micro:bit has two inputs that are ready to use: Button A and Button B, either side of the display, and now it's time to introduce those into your program.

To ensure you don't lose your 'Hello, World!' program, start a new project by clicking on the Projects menu and clicking on New, and then clicking on the Editor toggle to switch from

JavaScript Blocks mode to JavaScript mode. You get a fresh project with two lines of code and two blank lines already in place (see Figure 6-1 earlier in this chapter); any changes you make don't affect your previous project. To keep the two easy to find, remember to give this new project a name. Click on Untitled at the bottom of the screen and then type **Button Inputs** to rename the project.

Click on Line 4 of the program listing or move the cursor with the arrow keys on your keyboard so that Line 4 is highlighted. Then type the following:

```
input.
```

As before, as soon as you type the period, you'll see a list of all the possible functions from the Input category. Rather than choosing a function from the list, though, continue to type until Line 4 reads like this:

```
input.onButtonPressed(Button.A, (
```

When you reach the end of this line, the JavaScript Editor automatically adds a **)** symbol to the end of the line for you. This is another time-saving and bug-fixing feature of the JavaScript Editor. The code you are writing now, just like the loop you used to make the 'Hello, World!' program repeat its message, needs to be opened and closed using the bracket symbols (and). Because having an opening bracket without a matching closing bracket would be a bug, the JavaScript Editor automatically adds the closing bracket for you. This feature is known as *autocompletion*.

Just because the closing bracket is added automatically doesn't stop you from typing it. The JavaScript Editor is smart enough that if you type the closing bracket yourself, it will overwrite the one it added automatically rather than putting an extra closing bracket on the line—a bug which would bring your program to a halt just as quickly as having too many opening brackets.

Continue typing, ignoring the JavaScript Editor's autocompletion, until Line 4 reads like this:

```
input.onButtonPressed(Button.A, () => {})
```

Although this is a complete and valid line, it's not formatted in the best possible way for future clarity. Click between the curly brace symbols {} or use your arrow keys to move the input cursor there, and then press the Enter key. The JavaScript Editor automatically moves the last two symbols, }), to Line 6 and moves your cursor to Line 5 with a four-space indentation, ready for the next line of your program. Note, however, that any formatting you do to improve the readability of a JavaScript program will be lost if you use the Editor toggle to switch it into JavaScript Blocks mode and back again.

Again, as an alternative to typing, you can click on the Input category of the Blocks toolbox, click on the [input.onButtonPressed(Button.A, () => {})] block, or click on the autocompletion pop-up at any time during typing to have the code inserted automatically. In either of these cases, the code is formatted for you with the main portion on Line 4, a blank indented line on Line 5, and the closing section on Line 6 (see Figure 6-6).

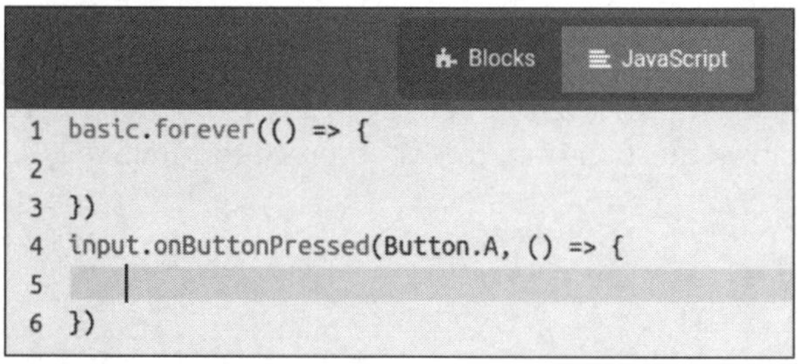

```
                              ✦ Blocks    ≣ JavaScript

1  basic.forever(() => {
2
3  })
4  input.onButtonPressed(Button.A, () => {
5     |
6  })
```

FIGURE 6-6: The input.onButtonPressed code block

The code you've just typed creates what is known in programming as an *event*; it works a little differently to the single-line instructions you've been using before. An event sits outside the main program, and any lines of code between its curly braces are ignored, but that changes when its *trigger* is activated. In the case of the input.onButtonPressed(Button.A, () => {}) event, that trigger is Button A being pressed. As soon as that happens—regardless of whatever else the program might be doing—any lines of code between the event's two curly braces begin to run.

At the moment, there is no code inside the curly braces. To fix that, put your cursor onto the blank, indented Line 5 if it isn't there already, and type this:

```
basic.showIcon(IconNames.Happy)
```

Feel free to click on the autocompletion pop-up at any time during the typing to save on keystrokes. This is particularly helpful when choosing the picture to display from a selection already drawn for you. After you type **IconNames.**, you're offered a list of possible icons from which to choose, which is easier than trying to remember what all of them are called!

Rather than scrolling a string as in 'Hello, World!', this function does exactly what its name suggests: tells the BBC micro:bit to load a picture, known as an *icon*, and display it on the screen. In this case, you've told the BBC micro:bit to show a happy face; other entries in the list of preprogrammed icons include a heart, a sad face, a butterfly, and a skull (see Figure 6-7).

FIGURE 6-7: Choosing an icon from the IconNames list

If you click on Button A in the simulator now, you see the effect of your program: a happy face appears on the screen. It stays there even when you're not holding the button down, too, because the program doesn't say otherwise. There's no instruction to display a different image or to clear the screen, so once the happy face is there, it remains until you reset the simulator or BBC micro:bit.

Events only run when their trigger is activated and run each time the trigger is activated. Press Button A once, and your event runs once. Press it again, and it runs again. Press it a third time, and it runs a third time, and so on.

Multiple Buttons

Being able to read a single button is great, but the BBC micro:bit has two. Thankfully, adding in the second button is as simple as repeating what you've already done. Move your cursor to the end of Line 6, either using the mouse or the arrow keys on the keyboard, and then press the Enter key to create a blank Line 7. Then type the following:

```
input.onButtonPressed(Button.B, () => {
    basic.showIcon(IconNames.Sad)
})
```

As before, you can use the JavaScript Editor's autocompletion feature to reduce the amount of typing you need to do. Alternatively, you can use your mouse—or Shift and the arrow keys on your keyboard. Highlight Lines 4 through 6 inclusive, press Ctrl+C (Cmd+C on macOS), move the cursor to Line 7—creating it by clicking on the end of Line 6 and pressing Enter if necessary—and press Ctrl+V (Cmd+V on macOS). Then change A for B and Happy for Sad in the duplicated code (see Figure 6-8).

```
      🔢 Blocks      {} JavaScript

1  basic.forever(() => {
2
3  })
4  input.onButtonPressed(Button.A, () => {
5      basic.showIcon(IconNames.Happy)
6  })
7  input.onButtonPressed(Button.B, () => {
8      basic.showIcon(IconNames.Sad)
9  })
```

FIGURE 6-8: The finished multibutton program

Click on Button A in the simulator—or press the physical button if you're running the program on a real BBC micro:bit—and the happy face appears as before. Click Button B, though, and you get a sad face; your program can read either button and run different events depending on which is pressed.

There's a third option for reading buttons, too: reading both at the same time. A `Button.AB` trigger in an event block only runs when both buttons are pressed at the same time, allowing you to have up to three inputs from two buttons.

As an experiment, try adding an event to your program, changing it to a `Button.AB` block, and picking a different icon to display. On a real BBC micro:bit, hold down both Button A and Button B to make the new block run; on the simulator, where you can't click on both buttons at the same time, you see an A+B button appear just below Button B. Pressing the A+B button is the equivalent of holding down both buttons.

Program 3: Touch Inputs

Although its two buttons give the BBC micro:bit up to three possible inputs, that might not be enough for your needs. This is where the input-output pins, the numbered contacts on the copper strip at the bottom of the BBC micro:bit, come into play. The BBC micro:bit can detect when you touch one of them with your finger and react accordingly—a fun alternative to the physical buttons and a great way to add extra inputs to a program without the need to buy and wire up more buttons.

As always, you need to start a new project by clicking on the Projects menu and clicking on New so you don't lose your work on the Button Inputs project. Click on Untitled at the bottom of the screen and then type **Touch Inputs** to name your new project.

Remember that you can always find your old projects by clicking on the Projects menu and looking under the My Stuff section. Projects are ordered according to when they were last modified, so your newest projects are first and your oldest projects last.   **TIP**

Reading a pin is a lot like reading a button. Start by moving your cursor to the blank Line 4; then type the following:

```
input.onPinPressed(TouchPin.P0, () => {

})
```

As before, you are given multiple pop-ups from the JavaScript Editor's autocompletion feature as you type; you can either click on these to have the code finished for you, or you can continue to type until the three lines—the middle line being blank—are finished. Alternatively, you can click on the Input category of the Blocks toolbox and click on the `[input.onPinPressed(TouchPin.P0, () => {})]` block.

Like the `input.onButtonPressed(Button.A, () => {})` block, `input.onPinPressed(TouchPin.P0, () => {})` represents an event. It doesn't need to be inside the `basic.forever(() => {})` loop, and any lines of code placed within it don't run until the event is triggered. Rather than being triggered by a button press, however, `input.onPinPressed(Button.A, () => {})` is triggered when you touch one of the pins—in this case, Pin 0.

Variables

The event now needs something to do. Move your cursor to the start of Line 1 and press Enter to move everything down by one line. Go back to Line 1, and then type the following to create a new *variable* (see Figure 6-9):

```
let touches = 0
```

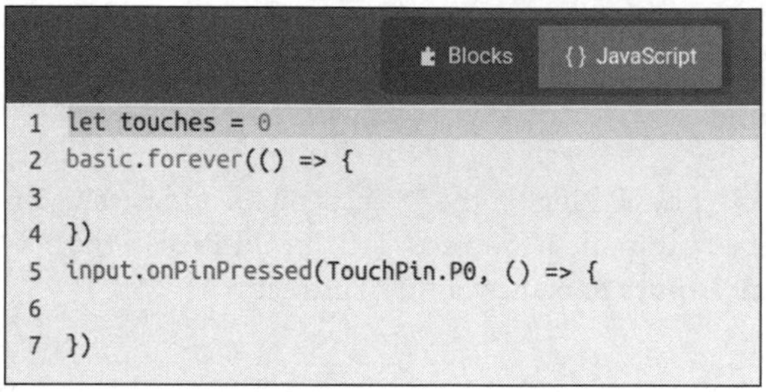

```
                                    Blocks    {} JavaScript

1  let touches = 0
2  basic.forever(() => {
3
4  })
5  input.onPinPressed(TouchPin.P0, () => {
6
7  })
```

FIGURE 6-9: Initialising a variable

Variables are a new concept. Like the name suggests, a variable is something which changes as the program is run. A variable can be almost anything: a number, a string of text, even the data required to draw a picture on the screen. The variable itself has two main properties: its *name* and its *data*. In the case of the variable you've just created, the variable name is `touches` and it has been set to store the value zero. The `let` instruction both creates the variable and *initialises* it. A variable must always be initialised before it is used, which is why the line `let touches = 0` is placed at the top of the program. Doing so creates a *global variable*, one which can be used by any other part of the program. Placing the `let touches = 0` line elsewhere in the program creates a *local variable* accessible only during that part of the program—for instance, within the `basic.forever(() => {})` loop.

> **TIP**   Variable names should always be as descriptive as possible, so their purpose within the program is obvious. There are restrictions on the names you can use, though: you can't use a name which is reserved for an instruction in the programming language you're using, they typically can't begin with a number, and they can't contain spaces or other symbols. To make variables with multiword names clearer to read without spaces, it's common practice to use 'camel case': 'Number of Fish', for example, becomes 'numberOfFish' in camel case; 'Age of User' becomes 'ageOfUser'. Camel case itself is often written as 'camelCase' to make its use more obvious.

Now that you have a variable, you need to do something with it. Move your cursor to the empty Line 6, within the `onPinPressed` event, and type the following (adding a four-space indent at the beginning if the JavaScript Editor has not done so for you):

```
touches += 1
```

> **TIP**   Like instructions, variable names are case sensitive. If you called your variable `touches` when you initialised it but `Touches` when you tried to change it, you get an error. Using the wrong case is a common source of bugs in JavaScript programs.

The program is now, technically, complete (see Figure 6-10), but it's not very useful. Each time you touch Pin 0, the `touches` variable increases, but there's no way to see that it's happening. The program needs an output, so press Enter to add a new, blank Line 7 and type the following:

```
basic.showNumber(touches)
```

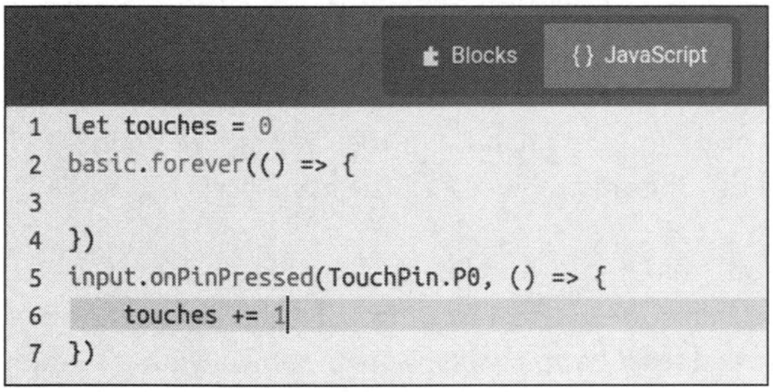

FIGURE 6-10: Modifying a variable

You've now created your first multiline event (see Figure 6-11). Each time the event is triggered, the lines of code inside run one at a time starting at the top and working downward until the end of the event, marked by the `})` on Line 8, is reached. By chaining lines of code like this, you can create programs as long and complex as you like.

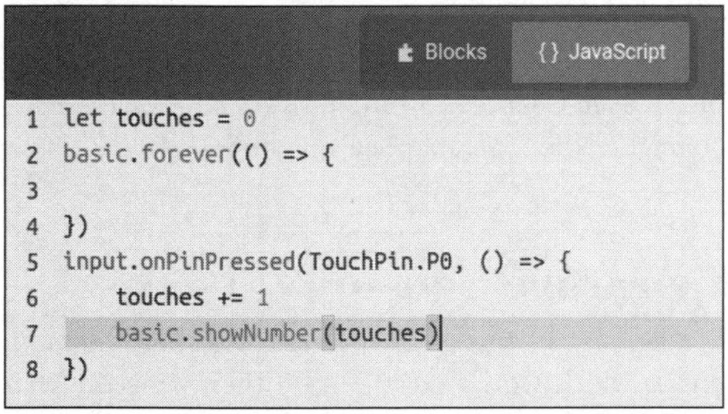

FIGURE 6-11: The finished program

The phrase `basic shownumber touches` doesn't make much sense in English, but if we break down Line 7, you can see what the computer sees: `basic` represents the category of instruction, whereas `showNumber` is the instruction, telling the BBC micro:bit to take a number—rather than a string—and show it on the display; `touches` isn't a number itself, but a variable which *contains* a number. Thus, when the line `basic.showNumber(touches)` runs, it prints to the display whatever number is stored in the `touches` variable.

If you're running the program in the simulator, click on Pin 0 now; the number 1 appears on the BBC micro:bit's display. Click on it again, and the number changes to 2; click on it again, and it changes to 3. The program continues to count the number of times you've clicked on the pin until you either press the Reset button—in which case the count resets to zero again—or disconnect the power supply.

If you're running the program on a physical BBC micro:bit, touching Pin 0 isn't enough. The BBC micro:bit's ability to sense physical touch works via an electrical property known as *resistance*, and it requires that the circuit—which is your body, in this instance—is *grounded*. Place the index finger of your right hand on the GND pin to the far right of the BBC micro:bit; then tap on Pin 0 with the index finger of your left hand. On the first tap, the number 1 appears; for each subsequent tap until the BBC micro:bit resets, the number increases.

This *resistive touch sensing* is available on Pin 0, Pin 1, and Pin 2 of the BBC micro:bit. Using the inputs available from the physical buttons—Button A, Button B, and Buttons A+B— plus the three pins, it's possible to read and react to up to six inputs in total in your program: the three button inputs plus touch on Pin 0, Pin 1, and Pin 2.

> **TIP** When you're counting the number of apples or pears you have, you typically start counting at one. When programming, though, you start at zero. That's why the BBC micro:bit's three main pins are labelled Pin 0, Pin 1, and Pin 2, rather than Pin 1, Pin 2, and Pin 3. The same goes for any other numbers you're working with in your program. If you're counting the number of times a loop runs, '9' in the program means it has run ten times—counting 0, 1, 2, 3, 4, 5, 6, 7, 8, 9, for a total of ten numbers—rather than the nine you might expect. It can take a while to get used to thinking from zero, so don't worry if you forget!

There's more on using the BBC micro:bit's pins in Chapter 10, 'Building Circuits', Chapter 11, 'Extending the BBC micro:bit', and Chapter 12, 'The Wearable BBC micro:bit'.

Program 4: The Temperature Sensor

The BBC micro:bit has more inputs than simple buttons and touch-sensing pins. The most simple of these is the temperature sensor, which works exactly like a thermometer. Read from the sensor, and it returns a temperature, measured in degrees Celsius.

Although the temperature sensor is enough to give you a rough idea of how hot or cold your surroundings are, though, it's not a precision instrument. It's built into the BBC micro:bit's processor and was originally designed for monitoring the processor's temperature rather than environmental temperatures. If the BBC micro:bit were a desktop computer, that would be a problem: desktop computer processors run tens of degrees hotter than their surroundings even when idle, requiring large metal heatsinks and fans to keep them cool. The BBC micro:bit's processor, though, runs close enough to the temperature of its surroundings—known as the *ambient temperature*—that it is typically accurate to within a degree or two.

If you're looking to get the most accurate temperature reading possible, it's important to keep your program simple. If your program is working the BBC micro:bit's processor hard—doing a lot of complicated sums, for instance—then the processor begins to warm up, throwing off your readings.

TIP

Start your program in the traditional way. Click on the Projects menu, click on New, and then click on the Editor toggle so you don't lose your work on the Touch Inputs project. Click on Untitled at the bottom of the screen, and then type **Temperature Sensor** to name your new project.

Move your cursor to the blank Line 2 and type the following line of code:

```
basic.showNumber(input.temperature())
```

As before, you can use the JavaScript Editor's autocompletion feature to cut down on the typing you need to do. This line represents a combination of two instructions (see Figure 6-12): `basic.showNumber`, which is used to show a number on the BBC micro:bit's display much like `basic.showString` printed your message in the 'Hello, World!' program, and `input.temperature`. This highlights a key feature of programming in JavaScript; rather than putting a number like 0 into the `basic.showNumber` instruction or even a variable like `touches` earlier in this chapter, you can put a whole instruction in there and have it print out the instruction's output—in this case, the reading from the BBC micro:bit's internal temperature sensor.

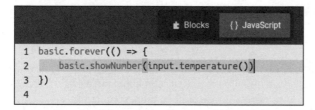

FIGURE 6-12: Reading from the temperature sensor

At this point the simulator springs into life and starts scrolling the current temperature across the BBC micro:bit's display. A slider also appears on the simulated BBC micro:bit, set to 21°C. Because the simulator doesn't have a real temperature sensor, this slider allows you to change the reported temperature from a pretend temperature sensor: slide the bar up to increase the reported temperature, and slide the bar down to decrease the reported temperature.

Formatting the Output

The program is now, technically, complete, but it's not easy to read. The numbers scroll across the screen continuously with no gap, making it hard to tell the difference between 21, 12, 22, or even 212121. To fix that, press Enter to add a new, blank Line 3 and type in the following, making sure there's a space between the first double-quote and the word `Celsius` (see Figure 6-13):

```
basic.showString(" Celsius")
```

```
                          ⚑ Blocks    {} JavaScript

1  basic.forever(() => {
2      basic.showNumber(input.temperature())
3      basic.showString(" Celsius")
4  })
5
```

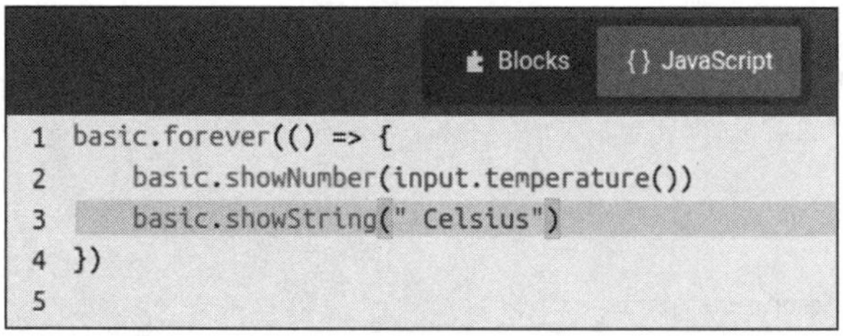

FIGURE 6-13: Formatting the program's output

The simulator updates immediately. Rather than printing the temperature reading continuously with no delay or gap in the number, it begins by printing the temperature followed by the word `Celsius` with a space between the two—which is the reason for including a space when you typed `Celsius` in Line 3.

If you're running your program on a physical BBC micro:bit, touch a metal surface to discharge any static electricity you may be holding (see Chapter 2, 'Getting Started with the BBC micro:bit'). Then carefully and gently place the tip of your finger over the small black square labelled PROCESSOR on the back of the BBC micro:bit. After a few seconds, you should see the temperature reading start to rise. Remove your finger, and after a few more seconds it should drop back down again.

WARNING Always be careful when touching components on the back of the BBC micro:bit. If you have not properly grounded yourself by touching a metal surface first, you could fire a static shock into the sensitive circuitry and potentially damage your BBC micro:bit beyond repair.

Program 5: The Compass Sensor

The BBC micro:bit's magnetic compass is one of the two sensors labelled on the back—the other being the accelerometer, detailed later in this chapter—and one of the most interesting. By reading the strength of the local magnetic field in three dimensions, it can work out the direction the BBC micro:bit is facing relative to magnetic north. This same technology is built into many smartphones and is used by mapping software to figure out which direction the user is facing.

The first step of this project should be obvious by now: click on the Projects menu, click on New, and then click on the Editor toggle to create a fresh, blank project. Click on Untitled at the bottom of the screen, and then type **Compass Sensor** to name your new project so that you can find it again in the future. There's nothing worse than a folder full of projects all called Untitled!

Begin by moving your cursor into the blank Line 2 and typing the following:

```
    basic.showString("Heading " +↵
 input.compassHeading())
```

> When a line of code would extend past the border of the page, a ↵ symbol is printed. When **TIP** you see this symbol, continue to type the code without pressing the Enter or Return keys. If you're not sure how a line of code should be entered, visit the website at **www.wiley.com/ go/bbcmicrobituserguide** to download plain-text versions of each program; these can then be used for reference or even simply copy and pasted directly into the editors.

The JavaScript Editor, as always, offers to automatically complete the line as you reach various key points. This is particularly handy after you've typed **input.**, as you'll be offered a list of all the BBC micro:bit inputs JavaScript can read (see Figure 6-14). You can either choose the compassHeading input from this list to complete the line of code or just ignore the pop-up and continue typing it out yourself.

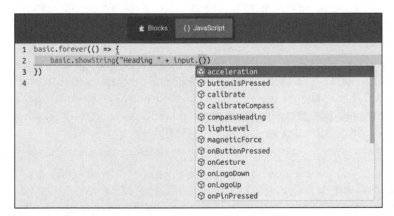

FIGURE 6-14: Choosing an input

The finished line is a little dense, so it requires a bit of unpacking. The `basic.showString` section should be familiar. This tells JavaScript to take a message, or string, and print it to the BBC micro:bit's display. In the Temperature Sensor program, you used this to print a helpful `Celsius` message after the actual sensor reading. Here it's doing the same thing—formatting the message printed to the BBC micro:bit's display—in a single line. The message begins with the word `Heading` and ends with the actual sensor reading.

The secret to this trick is surprisingly simple; it all lies in the + symbol (see Figure 6-15). This tells JavaScript to *join* two strings, which is exactly what it sounds like. If an instruction, like `basic.showString`, is expecting a string, using the + symbol allows you to give it two strings instead. In this case, the first string is the word `Heading` followed by a space, and the second string is the output of the `input.compassHeading` instruction. What took you two lines in the Temperature Sensor program takes only one in this program.

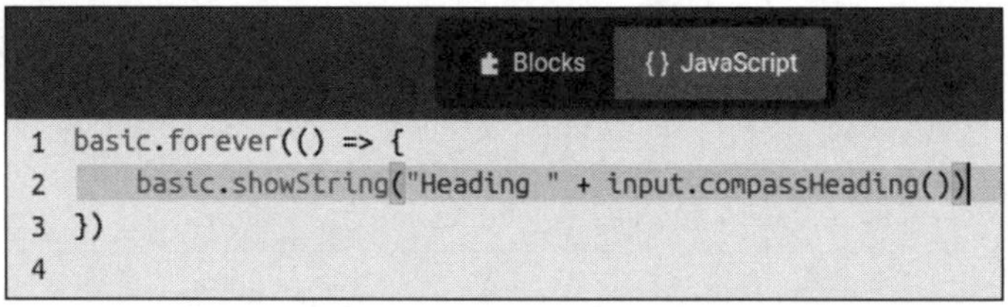

```
                              Blocks    {} JavaScript

1  basic.forever(() => {
2      basic.showString("Heading " + input.compassHeading())
3  })
4
```

FIGURE 6-15: The completed program

TIP   The + symbol as used for joining strings is distinct from the + symbol you'll use when working with actual numbers. Although they're identical to look at and typed using the same key on your keyboard, they work differently depending on whether you're working with strings or numbers. The strings `"1"` + `"1"` would print as `"11"`—literally `"1"` joined to `"1"`—rather than the `"2"` you would get if you were working with numbers.

If you're running the program in the simulator, you see the micro:bit logo has grown a point and the label 90°. Click and hold the mouse button on the logo and drag it around to twist it; as the logo twists, the compass heading changes with 0° (magnetic north) at the top. If you're running the program on a real BBC micro:bit, simply turn the BBC micro:bit itself—taking care not to dislodge the USB or battery cable.

TIP   If this is your first time using the compass, or if you've moved from one environment to another since the last time you used it, you may be asked to *calibrate* it. If the message `Draw a circle` appears on the display when you run your program, simply do as asked: rotate and tilt the BBC micro:bit so that the dot in the centre of the display rolls around the edge of the display and draws a circle. When the circle is complete, the compass is calibrated and the program runs normally. The accuracy of the calibration is very dependent on environmental factors. If you're calibrating the compass in an area surrounded by metal or near magnetic fields (such as on a metal desk, or next to a set of speakers), the compass will not calibrate correctly. Always try to calibrate the compass in an environment as close to that in which you'll be using it as possible for the highest accuracy.

One neat feature of using the + symbol to join strings is that you're not limited to joining just two: "1" + "2" + "3" + "4" gives you "1234", and you can keep adding strings using the + symbol without ever needing to write a second line of code. Try it yourself by modifying the existing program to print out the word degrees after the compass reading.

Program 6: The Accelerometer Sensor

The second of the two sensors labelled on the rear of the BBC micro:bit, alongside the compass, is the *accelerometer*. Where the compass is designed to measure magnetic fields and thus determine which direction the BBC micro:bit is facing, the accelerometer measures relative acceleration in three planes: X, Y, and Z. As well as being able to return actual sensor values, it can be used to detect different types of motion known as *gestures*—by far the simplest way to interact with the sensor in your programs.

The accelerometer is a powerful tool; besides being able to detect motion, it can be used to calculate the angle at which the BBC micro:bit is positioned by tracking the force of gravity pulling the BBC micro:bit toward the centre of the Earth—even when the BBC micro:bit is sat securely on a table or held in your hand. It's this feature of an accelerometer that a smartphone or tablet uses to detect when you've turned it from portrait to landscape orientation or vice versa so that it can automatically rotate the display to match.

To begin, click on the Projects menu, click on New, and then click on the Editor toggle to create a blank project. Click on Untitled at the bottom of the screen; then type **Accelerometer Sensor** to name your new project. If this part of the process seems repetitive, good. Keep repeating it until it's second nature to always create a new project and give it a name; then you don't run the risk of accidentally overwriting one of your previous programs or losing it in a sea of identically-named projects.

This project doesn't use the basic.forever(() => {}) loop. Although you've left these lines in place previously—because if there are no other lines of code inside, they are simply ignored when it's time to run the program—you can tidy the program listing by deleting them. Highlight the entire program with the mouse or keyboard, and then press the Backspace or Delete key to give you an entirely blank program listing with only one line.

A well-documented program is a good program. Although you won't be instructed to type comments during this chapter, as they're not required to make the program run, it's a good idea to get into the habit anyway. A part of the program which makes perfect sense today might seem confusing if you come back to the program a few months down the road and could be entirely impossible for someone else to understand without guidance. Comments don't slow a program down; use them liberally and you'll save yourself—and anyone else who uses your program—a great deal of trouble down the road.

TIP

To write a comment, simply type two slashes followed by the comment itself:

```
// Like this
```

A comment can be on a line by itself or come at the end of an existing line of code. Anything to the right of the two slashes is ignored by the JavaScript Editor and does not form part of the finished program, even if it's an otherwise valid line of code. If a comment is placed above a line of code, the JavaScript Editor will treat it as being attached to that line, and it will display as such if you use the Editor toggle to switch into JavaScript Blocks mode. If a comment doesn't have a line of code beneath it, though, it will be lost if you switch editor modes.

Start your program by typing the following three lines of code (see Figure 6-16) either entirely by hand or while allowing the JavaScript Editor's autocompletion feature to reduce the amount of typing required:

```
input.onGesture(Gesture.Shake, () => {
    basic.showIcon(IconName.Surprised)
})
```

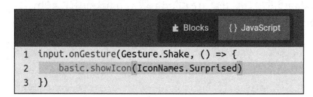

FIGURE 6-16: The onGesture event

The program is now ready for testing. When you added the `input.onGesture` instruction to your program, the simulator added a new control: a button marked SHAKE. Click on this button to simulate shaking the BBC micro:bit, and a shocked face appears on the display. Alternatively, move your mouse cursor over the BBC micro:bit board and wiggle your mouse to achieve the same shake effect.

The program works by using the accelerometer to look for one of a range of gestures—in this case, the shake gesture. If you delete the `.Shake` section from `Gesture.Shake` on Line 1 and then type a period, you see a list of the other gestures the accelerometer can track:

- **EightG**—The EightG—and ThreeG and SixG—gesture triggers only when the acceleration in any given direction exceeds eight times gravity, which is pretty fast.

- **FreeFall**—The free fall gesture triggers only when the BBC micro:bit has been dropped and is rushing toward the ground. This gesture is particularly useful when building flying robots or remote-controlled rockets.

- **LogoDown**—By tracking the pull of the Earth's gravity, the accelerometer can detect the orientation of the BBC micro:bit. This gesture triggers only when the BBC micro:bit is placed with the logo downward.

- **LogoUp**—The opposite of LogoDown, triggering only when the micro:bit logo is facing upward.

- **ScreenDown**—Tracked in the same way as LogoDown, the ScreenDown gesture triggers when the BBC micro:bit's display is facing downward.

- **ScreenUp**—The opposite of ScreenDown, this gesture triggers only when the BBC micro:bit's display is facing downward.

- **Shake**—Shake is activated by vigorously shaking the BBC micro:bit to and fro—but be careful you don't dislodge the micro-USB or battery cable or let go and send it flying across the room!

- **SixG**—The SixG gesture sits between the fast EightG and gentler ThreeG acceleration gestures.

- **ThreeG**—The ThreeG gesture works in the same way as the EightG gesture but triggers at a gentler rate of acceleration.

- **TiltLeft**—This gesture is based on the angle of the BBC micro:bit and triggers when the BBC micro:bit is tilted toward the user's left.

- **TiltRight**—The opposite of TiltLeft, the TiltRight gesture triggers when the BBC micro:bit is tilted to the user's right.

Delays

With a few changes, the program can become a lot more interactive. Move your cursor to the end of Line 2, if it isn't there already, and press Enter to add a new, blank Line 3. Then type the following:

```
basic.showIcon(IconNames.Asleep)
```

If you press the SHAKE button on the simulator now, you briefly see the Surprised icon before the Asleep icon takes over. If you were to continuously press the SHAKE button—the simulator equivalent of continuously shaking the BBC micro:bit without stopping—the face would change from the Surprised icon to the Asleep icon and back again over and over relatively quickly. That's not very realistic; it takes time to get back to sleep when you've been shaken awake!

To fix that, you're going to need to add a *delay* between the two `basic.showIcon` instructions. A delay does exactly what its name suggests: it pauses the program for a set length of time, delaying the execution of the next instruction. Move your cursor to the end of Line 2, press Enter to add a new, blank Line 3, and type the following (see Figure 6-17):

```
basic.pause(1000)
```

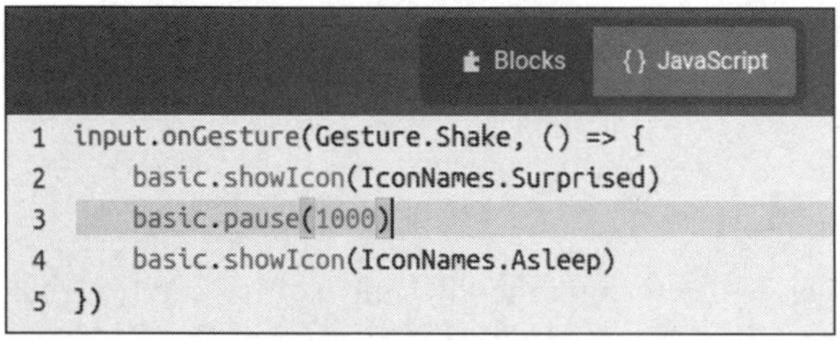

```
                              ⚫ Blocks    {} JavaScript

1   input.onGesture(Gesture.Shake, () => {
2       basic.showIcon(IconNames.Surprised)
3       basic.pause(1000)
4       basic.showIcon(IconNames.Asleep)
5   })
```

FIGURE 6-17: The completed program

This time, if you click the SHAKE button—or, if you're running the program on a physical BBC micro:bit, shake the BBC micro:bit—you see the Surprised icon longer before the Asleep icon takes over. Increase the number in the `basic.pause` instruction, and the Surprised face stays on the display for longer; decrease it and the Asleep face appears sooner.

The number between the brackets in the `basic.pause` instruction is a measure of how long the program should delay for in *milliseconds*, or thousandths of a second. A value of 1000, then, is equivalent to one second; 2000 would be two seconds, while 500 would be half a second. Because processors like the one powering the BBC micro:bit operate so much faster than humans, performing millions of operations every second, programs frequently need delays in them to allow the human operator to catch up.

As an experiment, see if you can extend the program still further by having the BBC micro:bit start on the Asleep icon rather than starting entirely blank until the first shake. Hint: code within your `input.onGesture` event only runs when the BBC micro:bit is shaken, so you need to put your code outside this event.

Reading Raw Accelerometer Data

The gesture system is by far the easiest way of working with the BBC micro:bit's accelerometer, but it's not the only way. You can also read the raw data into your program as numbers to handle yourself. Create a fresh project by clicking on the Projects menu, clicking on New, and then clicking on the Editor toggle. Click on Untitled at the bottom of the screen; then type **Accelerometer Sensor Data** to name your new project and keep it separate from the earlier Accelerometer Sensor gesture-triggered project.

> When a line of code would extend past the border of the page, a ↵ symbol is printed. When you see this symbol, continue to type the code without pressing the Enter or Return keys. If you're not sure how a line of code should be entered, visit the website at www.wiley.com/go/bbcmicrobituserguide to download plain-text versions of each program; these can then be used for reference or even simply copy and pasted directly into the editors.
>
> **TIP**

Start by moving your cursor into the blank Line 2, after the four-space indentation, and type the following either entirely by hand or using the JavaScript Editor's autocompletion feature:

```
    basic.showString("X:" +↵
  input.acceleration(Dimension.X))
```

This line is similar to the one you wrote in the Compass Sensor program. `basic.showString` tells the BBC micro:bit to print a string to the display, while the + symbol joins two strings: `"X:"` as a prefix followed by the output of the `input.acceleration(Dimension.X)` instruction. The simulator now starts scrolling the string "X:" followed by the current measurement from the accelerometer across the X axis. Try moving the simulated BBC micro:bit with your mouse cursor and watch the number change.

The X axis is only one of three axes that make up the accelerometer's three-dimensional measurements, though. To get at the others, click on the 2 at the start of Line 2 to highlight it and press Ctrl+C (Cmd+C on macOS) to copy it. Move your cursor to the end of Line 2, press Enter to add a new, blank Line 3, and then press Ctrl+V (Cmd+V on macOS) to paste the line of code you copied. Press Enter again to create a new Line 4, and then press Ctrl+V (Cmd+V on macOS) again to create another copy.

Your program is now displaying the accelerometer data three times, but it's the same X axis data each time. To correct that, you need to make four changes. In Line 3, change both X entries to Y; then in Line 4 change both X entries to Z to complete the program (see Figure 6-18).

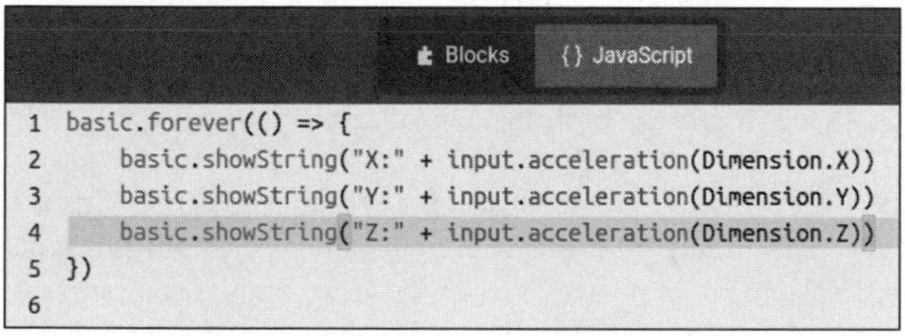

```
1  basic.forever(() => {
2      basic.showString("X:" + input.acceleration(Dimension.X))
3      basic.showString("Y:" + input.acceleration(Dimension.Y))
4      basic.showString("Z:" + input.acceleration(Dimension.Z))
5  })
6
```

FIGURE 6-18: The completed program

The BBC micro:bit now reads out all three of the axes, one after another: the X axis, followed by the Y axis, and then the Z axis before returning to the X axis again. Try moving the simulator to see the numbers change, or flash the program onto a real BBC micro:bit and see how the angle and position of the BBC micro:bit are reflected in the numbers—measured in *micro-gravities*—shown on the display. With the BBC micro:bit facing you, tilting it left will reduce the number reported on the X axis, while tilting it right will increase it. Tilting it so the top is closer to you will reduce the number reported on the Y axis, while tilting it so the top is further away will increase it. Finally, lifting the BBC micro:bit up will increase the number reported on the Z axis, while lowering it down will reduce it (see Figure 6-19).

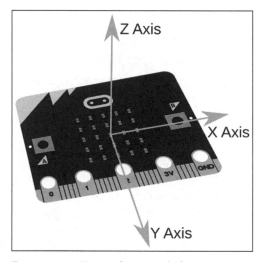

FIGURE 6-19: The accelerometer's three axes

Program 7: The Fruit Catcher Game

The programs in this chapter have been relatively simple, doing only one thing at a time, as a means of introducing some of the base concepts you need to know when programming the BBC micro:bit in JavaScript. This final program, though, is considerably more complex: a game which has the player—in the form of a single glowing pixel at the bottom of the display—attempt to catch ever-faster falling fruit—another pixel, starting at the top of the display and falling downward—before it hits the bottom of the display.

This program uses an instruction category not previously seen: the Game category. This is a category of instructions designed specifically for creating simple games, with built-in tools for keeping track of the score, displaying a 'game over' screen, creating single-pixel *sprites*—objects which can be moved around the display—and detecting when a sprite has collided with another sprite or the edge of the display.

The Fruit Catcher game is relatively simple but still quite a hefty program. If you've been working through this chapter nonstop, now might be a good time to take a break away from the screen so you can come back refreshed.

As ever, the game starts the same way as any program: creating a new project by clicking on the Projects menu, clicking New, clicking the Editor toggle, and giving it a name by clicking on Untitled and typing `Fruit Catcher`. You need the `basic.forever(() => {})` loop, so don't delete the existing line in the program listing. If you already deleted it before reading all of this paragraph, just click the Undo button to bring it back or type it back in by hand.

If you'd prefer to type the program out in one go and then read about what each line does afterward, you can find a full copy of the program code in Appendix B, 'JavaScript Recipes'. **TIP**

The Setup

The game should start by initialising some of the variables we need: a variable for adjusting the speed of the game called 'delay', a sprite for the player character, and another for the fruit. Move your cursor to the beginning of Line 1, press Enter to move everything down, and then move your cursor back to the beginning of Line 1 and type the following three lines of code:

```
let delay = 1000
let fruit: game.LedSprite = null
let player: game.LedSprite = game.createSprite(2, 4)
```

The first line sets the game's timing delay in milliseconds, demonstrating an important feature of variable initialisation: you can initialise a variable with whatever data you require, not just zero as in previous programs. The `delay` initialised in this line controls how fast the game is when it starts; if you'd like a challenge, you can make it lower; to make the game easier, make it higher. The second line initialises the `fruit` variable as a sprite using the instructions provided by the Game instructions category, while the third line does the same for the player variable—giving you the two objects the game needs: a fruit to fall and a player to catch it.

The third line has more to it than the second, however. Where the second line doesn't actually create the fruit sprite—as indicated by the word `null` at the end where the sprite's location would normally go—the third line tells JavaScript to create the sprite itself at a particular location on the BBC micro:bit's display. The numbers in the `game.createSprite(2, 4)` part of the instruction are important: they control exactly where the sprite appears on the BBC micro:bit's display. Every pixel on the display has a location on the horizontal X axis and the vertical Y axis. Figure 6-20 demonstrates this, giving the X and Y coordinates as X,Y for each of the 25 on-screen pixels.

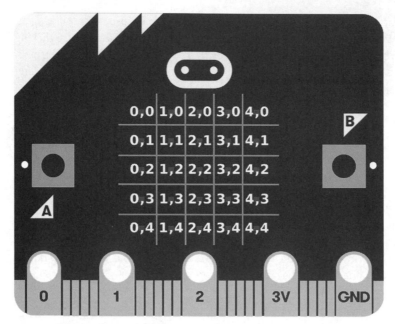

FIGURE 6-20: The BBC micro:bit Display Coordinates

By telling the `game.createSprite` instruction to create the player sprite at position 2 on the X axis and 4 on the Y axis—coordinate 2,4 in Figure 6-20—the sprite gets created in the middle column of the bottom row of the display.

Next, you need to initialise the game's score at zero. You could, if you wanted, cheat at this point and have the game start on a score of 10, or 100, or any other number you like, but where's the fun in that? Move your cursor to the end of Line 3, if it isn't there already, press Enter, and type the following line of code:

```
game.setScore(0)
```

With Line 4 now in place and a starting score initialised, the setup portion of your game is complete (see Figure 6-21).

```
                                      ✿ Blocks    {} JavaScript
1  let delay = 1000
2  let fruit: game.LedSprite = null
3  let player: game.LedSprite = game.createSprite(2, 4)
4  game.setScore(0)
5
```

FIGURE 6-21: The finished setup portion

The Main Program Loop

The next stage of the process is to make the game itself. Move your cursor to the end of Line 4, if it isn't there already, and press Enter to insert a blank Line 5. Line 6 should be the start of the `basic.forever` loop that was already in the program listing when you started the project, so move your cursor to Line 7 after the automatically-inserted four-space indentation and type the following:

```
fruit = game.createSprite(Math.random(5), 0)
```

The blank Line 5 is in your program for a reason: adding spaces between sections of code **TIP** makes it easier to see where one section ends and another begins. Blank lines, like comments, don't slow down your program or make the finished program any bigger, and they make it considerably easier for someone else to understand—or yourself, when you open the project in a few months' time and try to remember which section is which! Remember, though, that switching between JavaScript Blocks and JavaScript modes using the Editor toggle will remove formatting, changing the line numbers. Keep in JavaScript mode to avoid this happening to your program.

If the fruit appeared in the same place every time you played the game, it wouldn't be much of a game. Where the player sprite's starting position was set back in Line 3, Line 7 uses the `Math.random` instruction to find a random starting position every time a new fruit sprite is created.

The `Math.random` instruction does exactly what it says: picks a random number starting at zero and up to the range indicated by the number in brackets, which in this case is five. Because in programming numbers begin at zero, a range of five gives 0, 1, 2, 3, and 4—five numbers in total—as the possible output of the `Math.random` instruction. Helpfully, that range of five numbers maps perfectly to the X or Y axis of the display. That's how it's used here: when the game creates the fruit sprite, it positions it at the top (Y: 0) of the display but chooses a random location in that top row (X: 0 to 4, depending on what the random number generator picks).

At the rate the BBC micro:bit processor runs through blocks, a game where the fruit falls as quickly as possible would be many millions of times faster than a human could play; you'd see nothing but the `game over` screen. To fix that, put your cursor at the end of Line 7 if it isn't there already, press Enter, and type the following:

```
basic.pause(delay)
```

This tells the BBC micro:bit to pause for the number of milliseconds held in the variable `delay`, which you initialised back at Line 1 with a value of 1,000 for a one-second delay. At this point, the program should look like Figure 6-22.

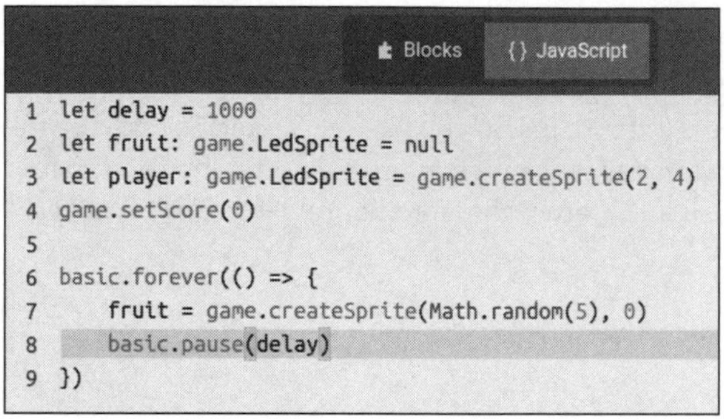

```
       ▲ Blocks    {} JavaScript
1  let delay = 1000
2  let fruit: game.LedSprite = null
3  let player: game.LedSprite = game.createSprite(2, 4)
4  game.setScore(0)
5
6  basic.forever(() => {
7      fruit = game.createSprite(Math.random(5), 0)
8      basic.pause(delay)
9  })
```

FIGURE 6-22: The beginnings of the main program

The Conditional Loops

So far, you've only worked with a single type of loop: the infinite loop, which runs forever. A more powerful type of loop is the *conditional loop*, which tests for a particular condition—such as whether a variable is equal to another variable—and runs only when the condition is true.

Move your cursor to the end of Line 8, if it isn't there already. Then press Enter and type the following:

```
while (fruit.get(LedSpriteProperty.Y) < 4) {
```

The JavaScript Editor, as always, attempts to help through autocompletion, including adding a closing curly brace (}) to the end of the line you're typing. Press Enter when you've typed the line above, and it automatically moves the closing curly brace down to Line 11 and leaves you on Line 10 with a new level of indentation: eight spaces instead of four, which indicates that the block of code you're typing in next sits under the conditional while loop you just created.

This line opens a while loop, a conditional loop that only runs when the condition it is testing is true, and retrieves the LedSpriteProperty.Y of the fruit variable—the location on the display's Y axis of the fruit sprite you created earlier. It then takes this value, which will be between zero for the top row of the display and four for the bottom row of the display, and checks to see if it is less than (<) four (see Figure 6-23). If, and only if, it is—which is to say that the fruit sprite hasn't yet reached the bottom of the screen—will the lines of code within the loop run.

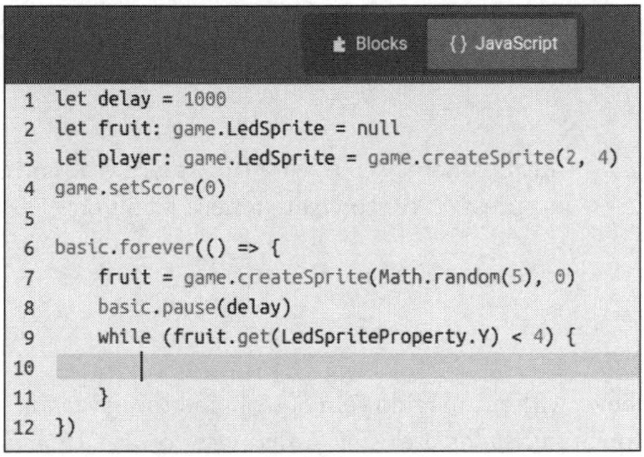

```
   Blocks   {} JavaScript

1  let delay = 1000
2  let fruit: game.LedSprite = null
3  let player: game.LedSprite = game.createSprite(2, 4)
4  game.setScore(0)
5
6  basic.forever(() => {
7      fruit = game.createSprite(Math.random(5), 0)
8      basic.pause(delay)
9      while (fruit.get(LedSpriteProperty.Y) < 4) {
10         |
11     }
12  })
```

FIGURE 6-23: The conditional loop

At the moment, though, the loop doesn't actually do anything. With your cursor on Line 10, after the eight-space indentation, type the following:

```
fruit.change(LedSpriteProperty.Y, 1)
```

With this line in place, every time this block of code runs—which is whenever the loop allows it to, which is only when the fruit's position on the Y axis of the display is less than four—the fruit's Y axis position is increased, or incremented, by one. The result: the fruit seems to be 'falling' down the display.

The falling process needs to be slowed down to make the game fun, though, so with your cursor at the end of Line 10 press Enter and type the following on Line 11:

```
basic.pause(delay)
```

Now the game waits for the number of milliseconds specified in the `delay` variable—which you initialised at 1000 for a delay of one second—before the fruit drops down to the next column.

The Conditional Statements

Loops aren't the only instructions to which conditions can be attached. A powerful programming tool is the if, then, else *conditional control flow*. Using an if, then, else control flow, you can test to see if a condition is true and then run a particular set of instructions, or you can run a different set of instructions if the condition is false. Unlike a conditional loop, a conditional control flow doesn't loop. Whatever instructions run as a result of the condition, they'll only run once unless you place them inside a loop within the conditional statement itself.

Move your cursor to the end of Line 12—which contains the closing curly brace that ends the conditional loop you created earlier—and press Enter to create a new, blank Line 13. Then type the following:

```
if (player.isTouching(fruit)) {
```

Press Enter after the end of the line above, without moving your cursor, and the JavaScript Editor's autocompletion function automatically adds a closing curly brace to Line 15 and places your cursor on Line 14 after eight spaces of indentation—again indicating that the lines you type next become part of the code run by the conditional statement if the condition turns out to be true.

This `player.isTouching(fruit)` instruction is the test for the conditional. It checks whether the player sprite, created at the start of the game, is touching the fruit sprite when the fruit sprite has reached the bottom of the screen. If the player sprite is touching the fruit sprite, the player has 'caught' the fruit; if not, the player has 'missed' the fruit.

If the player has caught the fruit, the player's score needs to be increased. With your cursor after the indentation on Line 14, where it will have been automatically placed, type the following:

```
game.addScore(1)
```

This conditional handles the case of the player successfully catching the fruit, but the game needs to be able to run a different instruction if the player misses. That's handled by adding an else case to the conditional. Press the Enter key to insert a new, blank Line 15; then type the following two lines:

```
} else {
    game.gameOver()
```

Notice that the JavaScript Editor automatically corrects the indentation, adjusting it from eight spaces to four, as you type the closing curly brace to end the if part of the conditional. Read as a whole, these instructions form the following sentence in plain English: if the player sprite variable is touching the `fruit` variable, add one to the score; if it is not touching the `fruit` variable, end the game and display the game over screen.

Code in the if portion of the conditional only runs if the conditional is true; code in the else portion of the conditional only runs if the conditional is false. No matter what happens, only one of the two blocks of code ever runs; under no circumstances does the program ever run both the if and the else statements on a single run-through of the loop.

Finally, you need two more lines of code to finish off the main portion of the game. Move your cursor to the end of Line 17, press Enter to insert a new, blank Line 18, and then type the following:

```
fruit.set(LedSpriteProperty.Brightness, 0)
```

Now when the player has caught the fruit sprite its brightness will be set to zero, hiding it from view. Place the last line of code for the main program by pressing Enter and typing the following on Line 19:

```
delay = delay - delay / 10
```

It's this line of code that is responsible for speeding the game up after every successful catch of the fruit. The first part of the code tells JavaScript to change the value of the `delay` variable to the outcome of a short equation; the second tells it to take something away from the current value of the `delay` variable; the last part tells it to take away the current value of the `delay` variable divided by 10. The result: the `delay` variable shrinks by 10 percent each time the fruit is caught, making the game 10 percent faster each round.

There's actually a neater way to decrease, or *decrement*, the value of a variable. Just as you used += in previous programs to increment a variable, -= decrements the variable. Thus, Line 19 should instead read:

TIP

```
delay -= delay / 10
```

Unfortunately, writing the line like this prevents you from being able to switch the project from JavaScript mode to JavaScript Blocks mode using the Editor toggle and back again without having changes made to the program. While Line 19 as written in this chapter isn't as neat as the above version, it offers better compatibility with the JavaScript and JavaScript Blocks Editors.

There are other impacts to switching from JavaScript to JavaScript Blocks mode and back again: you'll find that the JavaScript Blocks Editor makes changes to your program, including initialising all variables at zero and moving other parts of the program from the top to the bottom. These changes have no effect on the program's functionality, though: the program will run exactly the same, but the code may look a little different.

You've now finished the main program loop (see Figure 6-24). While the game may appear to be running in the simulator, there's a problem: you can't yet control the player sprite.

```javascript
   ● Blocks    {} JavaScript

1  let delay = 1000
2  let fruit: game.LedSprite = null
3  let player: game.LedSprite = game.createSprite(2, 4)
4  game.setScore(0)
5
6  basic.forever(() => {
7      fruit = game.createSprite(Math.random(5), 0)
8      basic.pause(delay)
9      while (fruit.get(LedSpriteProperty.Y) < 4) {
10         fruit.change(LedSpriteProperty.Y, 1)
11         basic.pause(delay)
12     }
13     if (player.isTouching(fruit)) {
14         game.addScore(1)
15     } else {
16         game.gameOver()
17     }
18     fruit.set(LedSpriteProperty.Brightness, 0)
19     delay = delay - delay / 10
20 })
21
```

FIGURE 6-24: The finished `basic.forever` loop

The Control Events

To catch the fruit, the player needs to be able to move left and right along the bottom of the BBC micro:bit's display. Handily, there's Button A to the left of the display and Button B to the right—just what you need for controlling the player sprite. Making these control events is effectively a reprise of the Button Inputs program you wrote earlier in the chapter, but with a little extra logic to help things along.

Move your cursor to the end of Line 20, press Enter twice to insert a blank line, and then type the following:

```
input.onButtonPressed(Button.A, () => {
```

As before, when you press Enter after typing in this line, the JavaScript Editor automatically adds a Line 22 with the closing curly brace and places your cursor on Line 21 with four spaces of indentation. Without moving the cursor, type the following:

```
if (player.get(LedSpriteProperty.X) > 0) {
```

Again, press Enter after typing in this line to have the JavaScript Editor automatically create the required closing curly brace on Line 25 and place your cursor on Line 24 with eight spaces of indentation ready for your next line of code.

The `if (player.get(LedSpriteProperty.X) > 0) {}` conditional exists to stop the player sprite from leaving the left edge of the display. It checks the player's current position on the X (horizontal) axis: if it's greater than or equal to 0, representing the left edge of the display, the player should be allowed to move further left; if not, the player is already at the leftmost edge and should not be allowed to move further left.

To actually make the player sprite move, make sure your cursor is still on Line 24 after the eight space indentation and type this:

```
player.change(LedSpriteProperty.X, -1)
```

Now when the player presses Button A but isn't already at the left edge of the screen, the player's position on the X axis is decreased by one—or decremented—moving the sprite one pixel to the left.

To save time in making the same event block for moving right, use your mouse cursor or the Shift and arrow keys on your keyboard to highlight Lines 22 through 26 inclusive and press Ctrl+C (Cmd+C on macOS) to copy them. Move your cursor to the end of Line 26, press Enter, and then press Ctrl+V (Cmd+V on macOS) to paste a copy of the control event.

To have the game watch for the player wanting to move right, make the following changes to your new event block. On Line 27, change `Button.B` to `Button.A`, on Line 28 change `< 0` to `> 4`, and on Line 29 change `-1` to `1`. The new control event should now read like this:

```
input.onButtonPressed(Button.B, () => {
    if (player.get(LedSpriteProperty.X) < 4) {
        player.change(LedSpriteProperty.X, 1)
    }
})
```

Now the game has full control. The first event watches for Button A being pressed, checks that the player has room to move to the left, and then moves the player sprite leftward one column. The second event watches for Button B being pressed, checks that the player has room to move to the right, and then moves the player sprite rightward one column. The game is complete and should look like Figure 6-25.

```
                                    Blocks    {} JavaScript
 1  let delay = 1000
 2  let fruit: game.LedSprite = null
 3  let player: game.LedSprite = game.createSprite(2, 4)
 4  game.setScore(0)
 5
 6  basic.forever(() => {
 7      fruit = game.createSprite(Math.random(5), 0)
 8      basic.pause(delay)
 9      while (fruit.get(LedSpriteProperty.Y) < 4) {
10          fruit.change(LedSpriteProperty.Y, 1)
11          basic.pause(delay)
12      }
13      if (player.isTouching(fruit)) {
14          game.addScore(1)
15      } else {
16          game.gameOver()
17      }
18      fruit.set(LedSpriteProperty.Brightness, 0)
19      delay = delay - delay / 10
20  })
21
22  input.onButtonPressed(Button.A, () => {
23      if (player.get(LedSpriteProperty.X) > 0) {
24          player.change(LedSpriteProperty.X, -1)
25      }
26  })
27  input.onButtonPressed(Button.B, () => {
28      if (player.get(LedSpriteProperty.X) < 4) {
29          player.change(LedSpriteProperty.X, 1)
30      }
31  })
```

FIGURE 6-25: The finished game

Before trying out your new game, check that the program has been entered correctly by comparing it to the version found at the back of this book in Appendix B, making any changes necessary. Once you're sure the program is correct, you can play the game on the simulator. Just click the Reset button, shaped like a pair of arrows chasing each other in a circle, and the game starts. Move the player sprite, at the bottom of the screen, with Button A and Button B, and try to catch the fruit sprite as it falls from the top to the bottom. Miss the fruit and it's game over, and you see a flashing animation, a GAME OVER message, and your score. To restart the game, just click the Reset button again; that works on both the simulator and a real BBC micro:bit.

To improve the game—and your programming skills—try making some modifications. Instead of controlling the player sprite with the buttons, how about using the accelerometer to watch for tilting and give the game motion control? Try adding multiple fruit which all fall down the screen at once to increase the challenge.

Further Steps

Now that you've finished the chapter on programming in JavaScript, why not try the same exercises in Python in the following chapter? If you skipped straight to this chapter, see how the same programs are implemented in a visual development environment in Chapter 5. If you'd prefer to stick with JavaScript, click on the Projects menu of the JavaScript Editor followed by the Make tab to bring up a range of additional projects to try, from a touch-sensitive love meter to a robot built from a milk carton.

Chapter 7

Python

In this chapter

- An introduction to programming in the Python Editor
- 'Hello, World!': Your first Python program
- Programs for reading from button inputs, touch inputs, and the temperature, accelerometer, and compass sensors
- Fruit Catcher: A simple game to program and play on your BBC micro:bit

THE PYTHON PROGRAMMING language, named not for the snake but for the comedy troupe Monty Python, started life in 1989 as a hobby project of developer Guido van Rossum. Since its official release in1991, Python has grown in popularity thanks to its flexibility and a beginner-friendly format free of the need for hundreds of curly braces or semicolons you find in rival languages. Instead, Python makes heavy use of *whitespace*—space or tab indentations—to control where blocks of code begin and end.

Technically speaking, the BBC micro:bit is not programmed in Python itself but a variant of the language designed specifically for the BBC micro:bit and other microcontroller-based development and educational boards: MicroPython. MicroPython is a version of Python created by Damien George with certain parts of the language's standard libraries which aren't required for microcontroller use stripped away and other libraries added to produce programs which run well on the BBC micro:bit and similar devices. The MicroPython code for the BBC micro:bit was contributed by volunteers from all over the world, led by Damien George, Nicholas Tollervey, and Carlos Pereira Atencio, in an effort to help people learn Python programming using the device.

As in Chapter 5, 'JavaScript Blocks', and Chapter 6, 'JavaScript', in this chapter you're introduced to the Python Editor and learn to write a series of programs that make the most of the

BBC micro:bit's capabilities. Finally, you're able to make a quick-fire game in which you're given the task of catching ever-faster falling fruits before they hit the ground. These programs are the same between Chapter 5, Chapter 6, and Chapter 7, allowing you to easily compare and contrast the three languages to find the one that suits you best.

Introducing the Python Editor

The Python Editor is an all-in-one development environment created specifically for the BBC micro:bit's Python language. Running entirely within the web browser of any modern computer, the Python Editor doesn't require you to install software on your computer. All you need to do to load the Python Editor is open the web browser of an Internet-connected computer, type **python.microbit.org** into the address bar, and press the Enter key (see Figure 7-1).

> **TIP** The Microbit Educational Foundation and the Python community are constantly working on improving the editor at `python.microbit.org`, which can mean that features will be added and the visual interface might change slightly. You can always get back to the version of the editor found in this chapter by visiting `python.microbit.org/v/1`.

FIGURE 7-1: The Python Editor

Following are the main features of the Python Editor:

- **The Download button**—Located at the top left and looking like an arrow pointing down to the micro:bit logo, this button downloads your program to your computer as a hex file ready to be flashed onto your BBC micro:bit (see Chapter 3, 'Programming the BBC Micro:bit').

- **The Save button**—Located next to the Download button and looking like a larger arrow pointing downward to an external hard drive, the Save button saves a copy of your program code to your computer as a `.py` (Python) text file. You can reload this code into the editor for later use, but you can't run it on the BBC micro:bit until it has been loaded into a hex file and flashed onto the BBC micro:bit.

- **The Load button**—Clicking on the Load button will allow you to browse the files stored on your computer for Python programs you have downloaded previously or written in other software. The Load button lets you load hex files from the Download button and py files from the Save button, restoring the program listing accordingly.

- **The Snippets button**—Looking like three cogs and located to the right of the Save button, the Snippets button gives you quick access to commonly used chunks of program code, such as functions, as a means of reducing the amount of typing you have to do.

- **The Help button**—Found next to the Snippets button on the top menu bar, the Help button should be your first port of call if you get stuck. Click the Help button to load a page introducing the features of the Python Editor and explaining how to use it to write your first program.

- **The Zoom buttons**—Smaller than the other buttons in the top menu bar, these buttons zoom in and out to make the program listing text larger or smaller. For more complex projects, zooming out allows you to see more of the project onscreen at once; if you find the program listing hard to read, zooming in makes things clearer.

- **The project name**—The first thing to do when you start a new project is to give it a descriptive name using the box at the right of the top menu bar. Always start by renaming a project from the default name to make it easier to find in the future.

- **The program listing**—Taking up the bulk of the screen and with line numbers to the left side, the program listing area contains the Python program on which you are currently working.

If you have already worked your way through the previous programming chapters, you may be wondering where the simulator has gone. The Python Editor doesn't have a simulator; programs should instead be downloaded using the Download button and flashed onto a physical BBC micro:bit, as described in Chapter 3.

TIP

If you don't yet have a physical BBC micro:bit but still want to experiment with programming one in Python, an unofficial simulator is available at **create.withcode.uk**. Any micro:bit Python program loaded into this site loads a simulator when you press Ctrl+Enter.

Program 1: 'Hello, World!'

'Hello, World!' is the traditional first program for any language. It's basic by design but offers a first glimpse of what a given language is like to program in with the bonus of a quick payoff in the form of a message proving that your program is alive and well.

Start by opening your browser and going to `python.microbit.org` to load the Python Editor. The Python Editor loads with a default program already typed in, as an introduction. Although this program is effectively a 'Hello, World!' program itself, start by deleting all the text currently in the program listing to begin with a completely blank slate. The quickest way to do this is to press Ctrl+A on the keyboard to select all the text, and then press either the Backspace or the Delete key.

'Hello, World!' is a simple program that requires just two lines of code. The first is required by all micro:bit Python programs, regardless of what the program is designed to do, and you use the same line in every program in this chapter. With all text deleted from the program listing and your cursor at the start of the now-blank Line 1, type the following:

```
from microbit import *
```

This is known as an *import* instruction; it tells Python that you will be using instructions found in the *library* called `microbit`. A Python library contains a collection of pre-written code which is made available as functions, and the `microbit` library's functions revolve around using the BBC micro:bit in your program. Without this line, which tells Python to add all the instructions from the `microbit` library into your program ready for you to use, you wouldn't be able to use the BBC micro:bit's screen, sensors, input-output pins, or any other feature.

The second line of your program is the one that actually does the work of printing out the "Hello, World!" message to the BBC micro:bit's display. With your cursor at the end of Line 1, press Enter to insert a new, blank Line 2 and type the following:

```
display.scroll('Hello, World!')
```

This line tells Python that you are using the `scroll` instruction from the `display` category of instructions and that you are passing it the message "Hello, World!" (see Figure 7-2). As the name suggests, this instruction takes the message you have written and scrolls it across the BBC micro:bit's display for you to read.

FIGURE 7-2: The `display.scroll` instruction in the program listing

TIP If you make a mistake at any point while writing your program, you can undo your last actions by pressing Ctrl+Z on your keyboard. If you accidentally undo something you meant to keep, press Ctrl+Y to undo the undo—known as a *redo*.

TIP In Python, everything is *case sensitive,* meaning it always has to be typed in the exact right mix of uppercase and lowercase. The instruction `display.scroll` is valid, but `Display.Scroll`, `DISPLAY.SCROLL`, `display.Scroll`, and any other variant simply does not work.

You may have noticed that the line of code you've just written uses different colours. This is known in programming as *syntax highlighting* and helps you see any mistakes you may have made at a glance. The instruction itself, the `display.scroll` portion of the code, is displayed in an off-white colour; the message that the instruction is using, `'Hello, World!'`, is displayed in purple.

Try deleting the last single quote from Line 2, so it reads like this:

```
display.scroll('Hello, World!)
```

Notice how the colours have changed. The last bracket, which was off-white before, is now purple, indicating that the Python Editor thinks it's part of the string (see Figure 7-3). Without the last single quote to tell the Python Editor where the string ends, though, the program no longer works; you've introduced a *bug* into the program.

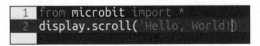

FIGURE 7-3: The Python Editor highlighting a code bug

The Python Editor's debugging capabilities are relatively basic. The Editor's key tool is the syntax highlighting: by looking at the colours, it's possible to see where common errors like strings that have not been closed are in your program. Even with the error you've introduced onto Line 2, though, the Python Editor still downloads your program when you click the Download button. You don't see error messages until you flash the program onto your BBC micro:bit, at which point you see any error message the compiler created scroll across the display—in this case, instead of the "Hello, World!" message you were expecting. This scrolling error message always includes a reference to the line number on which the error was found, allowing you to go back to your program, find, and fix the problem.

Put the single quote back into Line 2 so it reads like this:

```
display.scroll('Hello, World!')
```

The syntax highlighting changes the bracket back to its original colour to indicate that the bug is now fixed. Click the Download button to compile and save your program as `micro-bit.hex`, connect your BBC micro:bit to your computer via the micro-USB port, and then drag the `microbit.hex` file from your Downloads folder to the BBC micro:bit's removable drive as described in Chapter 3. Once the flashing process has finished, the program begins to run and the message "Hello, World!" begins to scroll across your BBC micro:bit's display (see Figure 7-4). Feel free to experiment at this point: change Line 2 to scroll a message of your own, such as your name or your favourite colour, and then click the Download button again to download an updated copy of your program to flash onto the BBC micro:bit and replace the original. If you haven't deleted or moved the original microbit.hex file from your Downloads folder, the new copy is called microbit (1).hex; if you click the Download button again, the next copy is called microbit (2).hex, and so on for each time you download a new version of the program.

TIP   In programming, messages like this are called *strings* and can be made up of letters, numbers, and some symbols. Although a string is great for displaying messages on a screen, you're limited in how you can work with it. The string "1", for example, can't be used in a mathematical equation because although it looks like a number to us, it's simply a string to the programming language. For mathematical equations, there are other types: *integers* are whole numbers, *floats* are numbers with decimal places, and *Booleans* are logic elements that can be True or False—equivalent to 1 and 0, respectively.

When you're happy with your program, make sure to click the Save button to save a copy of the program listing to your computer. Find this file, which will be called `microbit.py`, in your Downloads folder. Move it to a new location, such as your computer's Documents folder, and give it a more descriptive name, like `helloworld.py`, to make it easier to find in the future.

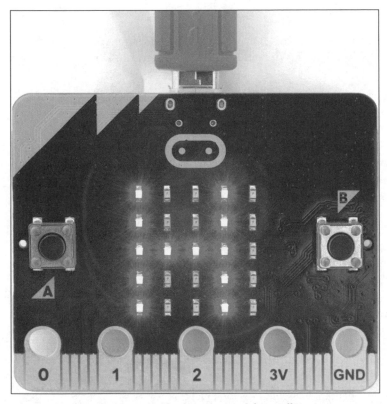

FIGURE 7-4: The BBC micro:bit displaying part of the scrolling message

It's important both to save a copy of your program listing using the Save button and to move it somewhere safe with a descriptive name. The Python Editor doesn't save any of your programs itself, meaning that if you close your web browser and come back to the Python Editor later, your program will have been erased and replaced with the default starting program. Renaming your program files, meanwhile, stops you from needing to sort through a dozen files all named `microbit` to find the program you need.

To load a saved program into the Python Editor, simply find the `.py` file on your computer, click and hold the mouse button over it, and drag the file into the program listing section of the Python Editor or click the Python Editor's Load button and browse to the file you want to load. Note that you can also do this with the hex file you get by clicking the Download button, as each hex file also contains a copy of the program listing.

TIP

Whatever message you decide to scroll, you'll notice it only scrolls once. You can make the message appear again if you reset the BBC micro:bit by pressing the Reset button on its rear, but if you don't want to have to keep doing that, you need a method of telling the program to repeat.

Loops

Move your cursor to the end of Line 1, and then press the Enter key to move Line 2 down to Line 3 and create a new, blank Line 2. With your cursor at the start of the new Line 2, type the following:

```
while True:
```

This creates a means of having parts of your program repeat, but at the moment Python doesn't understand which parts you actually want to repeat. To fix that, move your cursor to the start of Line 3 and press the spacebar four times to insert four spaces. When you're finished, the line will be four spaces inward compared to Lines 1 and 2.

This is known as *indentation*, and it is a way to easily see how lines of code in a program *nest* within each other. Indentation is usually handled automatically by the Python Editor when required, but not when editing existing lines of code. In many other languages, indentation exists only to make the code easier for the programmer to read—the code could be entirely flat against the left edge of the program listing and still work perfectly—but in Python indentation is a requirement. The level of indentation on each line of code in the program listing shows Python where in the program it belongs. If Python is expecting indentations and there aren't any, the program won't run properly; equally, if there are indentations where there shouldn't be, the program won't work either.

With your code properly indented, it should now look like this:

```
from microbit import *
while True:
    display.scroll('Hello, World!')
```

Click the Download button to download your modified program. Then find it in your Downloads folder and drag it to the MICROBIT drive to flash it. When the flashing process has finished, you see the familiar "Hello, World!" message, but this time it repeats forever (see Figure 7-5)—or at least until you unplug the BBC micro:bit.

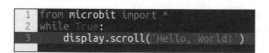

FIGURE 7-5: Looping the display.scroll message

In programming, this is known as a *loop*. When the program reaches the end of its instructions—in this case, when it has finished scrolling the message from the display. scroll('Hello, World!') instruction you moved to Line 3—it returns to the beginning of the loop and starts all over again. Loops in Python are *conditional*, which means they always loop if a test carried out at the start of the loop comes out at True. The loop you created in Line 2 has a simple test. It checks to see if True is, in fact, true. Because True is always true—which is why it's called True—the test always succeeds, and the loop continues forever; in programming, this is known as an *infinite loop*.

All loops need to be opened and closed, just like all strings need to be between paired double quotes. The : symbol at the end of Line 2 indicates the start of the loop, and any line that follows this and has a four-space indentation will become a part of the loop and run forever. If your program was longer and had lines of code that should not be part of the loop, they would be placed at the end and without the four-space indentation to indicate that they are not part of the four-space-indented loop code block.

TIP

Program 2: Button Inputs

'Hello, World!' demonstrated how to get an output from the BBC micro:bit via the screen, but there's another key feature of most programs: inputs. Thankfully, the BBC micro:bit has two inputs that are ready to use: Button A and Button B, either side of the display, and now it's time to introduce those into your program.

To ensure you don't lose your 'Hello, World!' program, make sure you've saved it using the Save button, moved the `.py` file to a safe place like your Documents folder, and given it a descriptive name like `helloworld.py`. Once you've done this, you are free to delete the current program listing by pressing Ctrl+A on your keyboard to select the text and then the Delete or Backspace key to delete it, leaving you with a fresh and blank program listing.

Begin by typing the first line of any BBC micro:bit Python program, importing the `micro-bit` library:

```
from microbit import *
```

Next, begin an infinite loop—just like the one you created in 'Hello, World!' to make your message scroll across the screen without stopping—by typing the following on Line 2, after pressing Enter at the end of Line 1:

```
while True:
```

When you press Enter at the end of this line, you might notice that your cursor doesn't go to the beginning of Line 3. Instead, it's positioned in Line 3 after a four-space indentation. This is known as *auto-indentation* and is a time-saving feature of the Python Editor: Because you opened an infinite loop on Line 2, and Python requires that the contents of that loop are indented by four spaces, the Python Editor has automatically inserted that indentation for you.

Auto-indentation works on multiple levels. If you open another loop within the current loop, the next line is automatically indented by eight spaces: four for the initial loop and another four for the inner loop. Put a loop in the loop in the loop, and the next line is indented by 12 spaces, 4 for each level of the loop.

Although auto-indentation creates indentation for you, it's up to you to delete the indentation when you no longer require it. The Python Editor doesn't know when you've finished writing code for your loop, only that you had started; when you come to the last line of your loop, you need to manually delete the indentation from the next line using the Backspace key on your keyboard. One press of the Backspace key when your cursor is to the right of the indentation deletes one level, or four spaces. Remember that Python uses this indentation to tell which lines of code are part of a loop and which lines of code aren't, so it's important to get the indentation right.

Type the following on Line 3, after the four-space indentation the Python Editor created for you:

```
if button_a.is_pressed():
```

When you press Enter at the end of this line, Python auto-indents the next line by eight spaces in total: four for the original infinite loop opened on Line 2, and another four for the line you've just created. This line is known as a *conditional*, or more specifically an *if statement*. An if statement in Python works just like an if statement would in English: 'if we're out of milk, go to the shop' tells someone to go to the shop but only *if* the milk has run out; if the milk hasn't run out, there's no need to go to the shop.

In the case of code lines you've just typed, the condition is simple: `if button_a.is_pressed()` checks to see if Button A on the BBC micro:bit is being pressed. If it is, the lines following the colon (:) at the end of Line 3 and featuring an eight-space indentation run; if it isn't being pressed, they are ignored.

It can be difficult at first to keep track of what code runs where, but it's here that the indentation really helps. If lines of code are lined up at the left, they're nested together and run one after another.

At the moment, though, there are no lines of code for the `if` statement to run regardless of whether or not the statement is true. On Line 4, to the right of the eight-character indentation, type the following:

```
display.show(Image.HAPPY)
```

Like the display.scroll instruction you used in the 'Hello, World!' program earlier, this instruction comes from the display category. Rather than scrolling a string across the BBC micro:bit's display, though, it displays an image chosen from a preprogrammed selection—in this case, a happy, smiling face. At this point, you should have four lines of code: the first two flush to the right of the program listing, the next line indented by four spaces, and the next indented by eight (see Figure 7-6).

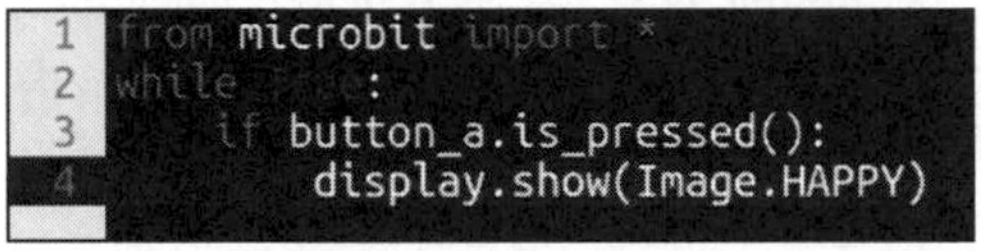

```
1  from microbit import *
2  while    :
3      if button_a.is_pressed():
4          display.show(Image.HAPPY)
```

FIGURE 7-6: The `if button_a.is_pressed()` code block

Click the Download button, flash the resulting file to the BBC micro:bit, and press Button A. You see your smiley face appear on the screen and stay there. If you press the BBC micro:bit's Reset button, the smiley face disappears until you press Button A again. You're not limited to a smiley face, either. You can choose from the built-in images by picking one of the following:

ANGRY, ARROW_N, ARROW_NE, ARROW_E, ARROW_SE, ARROW_S, ARROW_SW, ARROW_W, ARROW_NW, ASLEEP, BUTTERFLY, CHESSBOARD, CLOCK12, CLOCK11, CLOCK10, CLOCK9, CLOCK8, CLOCK7, CLOCK6, CLOCK5, CLOCK4, CLOCK3, CLOCK2, CLOCK1, CONFUSED, COW, DIAMOND, DIAMOND_SMALL, DUCK, FABULOUS, GHOST, GIRAFFE, HAPPY, HEART, HEART_SMALL, HOUSE, MEH, MUSIC_CROTCHET, MUSIC_QUAVER, MUSIC_QUAVERS, NO, PACMAN, PITCHFORK, RABBIT, ROLLERSKATE, SAD, SILLY, SKULL, SMILE, SNAKE, SQUARE, SQUARE_SMALL, STICKFIGURE, SURPRISED, SWORD, TARGET, TORTOISE, TRIANGLE, TRIANGLE_LEFT, TSHIRT, UMBRELLA, XMAS, YES.

Simply change the name after the `Image.` part of Line 4—remember that Python is case-sensitive, so the image name must be in all capitals as in the previous list, and keeping the full-stop character in place—to an image of your choice to modify what the program displays (see Figure 7-7).

```
1  from microbit import *
2  while True:
3      if button_a.is_pressed():
4          display.show(Image.PACMAN)
```

FIGURE 7-7: Choosing a different image to display

Conditional statements, like the `if` statement you created on Line 3, only run when their condition is true. They also only run once without looping; if you hadn't created an infinite loop on Line 2, the `if` statement on Line 3 would run through its code once, and then the BBC micro:bit would stop listening for button presses.

Multiple Buttons

Being able to read a single button is great, but the BBC micro:bit has two. Thankfully, adding the second button is as simple as repeating what you've already done: move your cursor to the end of Line 4, using either the mouse or the arrow keys on the keyboard, and then press the Enter key to create a blank Line 5. The Python Editor assumes that you are continuing the code for the `if` statement you created on Line 3 and automatically indents by eight spaces; press the Backspace key on the keyboard once to delete four of these spaces, leaving four, to indicate that the line you are about to type is not part of the `if` statement but is part of the infinite loop you created on Line 2. With that done, type the following two lines:

```
if button_b.is_pressed():
    display.show(Image.SAD)
```

As before, when you press the Enter key at the end of Line 5, Python automatically indents Line 6 to make it part of the second `if` statement. If you would like to reduce the amount of typing you need to do, you can use your mouse—or Shift and the arrow keys on your keyboard—to highlight Lines 3 and 4 inclusive, press Ctrl+C (the keyboard shortcut for Copy), and then move the cursor to Line 5—creating it by clicking on the end of Line 4 and pressing Enter if necessary—and press Ctrl+V (the keyboard shortcut for Paste), and then change A for B and HAPPY for SAD in the duplicated code. If you're pasting code like this, however, it can confuse the auto-indentation feature of the Python Editor. You'll find that Line 5 is indented by eight spaces when it should be indented by four. Move your cursor to the left of the word `if` on Line 5 and press the Backspace key once to delete the extra four spaces (see Figure 7-8). When pasting code into the Python Editor, always check that the indentation has been created correctly; remember that Python programs with incorrect indentation don't run properly.

```
1  from microbit import *
2  while True:
3      if button_a.is_pressed():
4          display.show(Image.HAPPY)
5      if button_b.is_pressed():
6          display.show(Image.SAD)
```

FIGURE 7-8: The finished multibutton program

Click the Download button and flash the resulting hex file onto the BBC micro:bit. Press Button A, and the happy face appears as before. Click Button B, though, and you get a sad face. Your program can read either button and run different events depending on which is pressed.

There's a third option for reading buttons, too: reading both at the same time. By chaining two conditions into a single `if` statement, it's possible to check if both buttons are pressed at the same time, allowing you to have up to three inputs from two buttons. The resulting `if` statement would be this:

```
    if button_a.is_pressed() and↵
button_b.is_pressed():
```

> **TIP**
>
> When a line of code would extend past the border of the page, a ↵ symbol is printed. When you see this symbol, continue to type the code without pressing the Enter or Return keys. If you're not sure how a line of code should be entered, visit the website at **www.wiley.com/ go/bbcmicrobituserguide** to download plain-text versions of each program; these can then be used for reference or even simply copy and pasted directly into the editors.

As an experiment, try adding this third `if` statement to your program and picking a different image to display. When you've flashed the modified program to your BBC micro:bit, hold down both Button A and Button B to make the new image appear.

When you're happy with your Button Inputs program, click the Save button to save a copy of the program listing to your computer, find it in the Downloads folder, and move it somewhere safe before giving it a more descriptive name like `buttoninputs.py`.

Program 3: Touch Inputs

Although its two buttons give the BBC micro:bit up to three possible inputs, that might not be enough for your needs. This is where the input-output pins, the numbered contacts on the copper strip at the bottom of the BBC micro:bit, come into play: the BBC micro:bit can detect when you touch one of them with your finger and react accordingly—a fun alternative to the physical buttons and a great way to add extra inputs to a program without the need to buy and wire up more buttons.

If you haven't yet saved your previous program, do so now and then press Ctrl+A followed by the Delete or Backspace key on your keyboard to delete the old program listing and start afresh.

Reading a pin is a lot like reading a button. Begin with the import line so you can access the `microbit` library's functions:

```
from microbit import *
```

Remember that all BBC micro:bit Python programs need this line, without which you can't use any of the BBC micro:bit's features, including the display and the buttons.

Variables

Type the following on Line 2 to create a new *variable* (see Figure 7-9).

```
touches = 0
```

FIGURE 7-9: Initialising a variable

Variables are a new concept. Like the name suggests, a variable is something that changes as the program is run. A variable can be almost anything: a number, a string of text, even the data required to draw a picture on the screen. The variable itself has two main properties: its *name*, and its *data*. In the case of the variable you've just created, the variable name is `touches`, and it has been set to store the value zero. The `let` instruction both creates the variable and *initialises* it. A variable must always be initialised before it is used, which is why the line `touches = 0` is placed near the very top of the program, just under the library import line. Doing so creates a *global variable*, one that can be used by any other part of the program; placing the `touches = 0` line elsewhere in the program creates a *local variable* accessible only during that part of the program—for instance, within a `while True:` loop.

> **TIP**   Variable names should always be as descriptive as possible, so their purpose within the program is obvious. There are restrictions on the names you can use, though. You can't use a name that is reserved for an instruction in the programming language you're using, they typically can't begin with a number, and they can't contain spaces or other symbols. To make variables with multiword names clearer to read without spaces, it's common practice to use 'camel case': 'Number of Fish', for example, becomes 'numberOfFish' in camel case; 'Age of User' becomes 'ageOfUser'. Camel case itself is often written as 'camelCase' to make its use more obvious.

Now that you have a variable, you need to do something with it. Press Enter at the end of Line 2 to create a new blank Line 3, and then type the following (remembering that the Python Editor automatically indents Line 4 by four spaces when you press Enter at the end of Line 3):

```python
while True:
    if pin0.is_touched():
```

This `if` statement works in the same way as the ones you created in the Button Inputs program: it forms a conditional that only runs if the condition is true. In this case, the condition is checking whether Pin 0 is being touched. If Pin 0 is being touched, the indented code under the `if` statement runs; if it's not being touched, the code doesn't run.

The `if` statement needs something to do, so press Enter at the end of Line 4 to create a new blank Line 5, with the Python Editor automatically inserting an eight-space indent for you, and type the following:

```python
        touches += 1
```

Like instructions, variable names are case sensitive. If you called your variable `touches` when you initialised it but `Touches` when you tried to change it, you get an error. Using the wrong case is a common source of bugs in Python programs.

TIP

The program is now, technically, complete (see Figure 7-10), but it's not very useful. Each time you touch Pin 0, the `touches` variable increases, but there's no way to see that it's happening. The program needs an output, so press Enter to add a new, blank Line 6 with an automatic eight-space indent, and type the following:

```python
        display.scroll(str(touches))
```

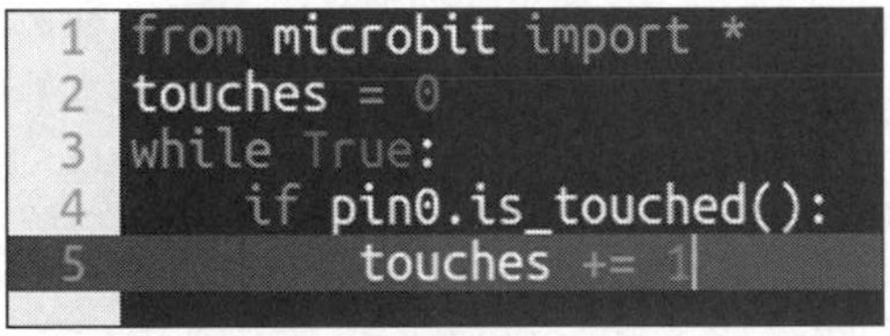

```python
1  from microbit import *
2  touches = 0
3  while True:
4      if pin0.is_touched():
5          touches += 1
```

FIGURE 7-10: Modifying a variable

You've now created your first multiline `if` statement (see Figure 7-11): each time the `if` statement is found to be true as the program runs through its infinite loop, the lines of code inside run one at a time starting at the top and working downward until the end of the eight-space indentation is reached. By chaining lines of code like this, you can create programs as long and complex as you like.

```
1  from microbit import *
2  touches = 0
3  while True:
4      if pin0.is_touched():
5          touches += 1
6          display.scroll(str(touches))
```

FIGURE 7-11: The finished program

The phrase `display scroll str touches` doesn't make much sense in English, but if we break down Line 6, you can see what the computer sees: `display` represents the category of instruction while `scroll` is the instruction, telling the BBC micro:bit to scroll a message on the display, just like in the 'Hello, World!' program. This instruction expects a string, but the `touches` variable you created earlier contains a whole number, or an *integer*. An instruction that expects to work on a string gets confused if you pass it an integer—and vice versa—even though the two look identical to humans. Before the `display.scroll` instruction can print the contents of the `touches` variable to the screen, it needs to be converted into a string. That's exactly what the `str` part of the instruction does: anything between its two brackets is converted into a string from whatever type it was before. Thus, when the line `display.scroll(str(touches))` runs, it prints to the display whatever number is stored in the `touches` variable.

Click on the Download button to compile and download your program, and then flash it to the BBC micro:bit. When the program begins to run, you find that touching Pin 0 isn't enough. The BBC micro:bit's ability to sense physical touch works via an electrical property known as *resistance*, and it requires that the circuit—which is your body, in this instance—is *grounded*. Place the index finger of your right hand on the GND pin to the far right of the BBC micro:bit; then tap on Pin 0 with the index finger of your left hand. On the first tap, the number 1 appears; for each subsequent tap until the BBC micro:bit resets, the number increases.

This *resistive touch sensing* is available on Pin 0, Pin 1, and Pin 2 of the BBC micro:bit. Using the inputs available from the physical buttons—Button A, Button B, and Buttons A+B—plus the three pins, it's possible to read and react to up to six inputs in total in your program: the three button inputs plus touch on Pin 0, Pin 1, and Pin 2.

TIP

When you're counting the number of apples or pears you have, you typically start counting at one. When programming, though, you start at zero. That's why the BBC micro:bit's three main pins are labelled Pin 0, Pin 1, and Pin 2, rather than Pin 1, Pin 2, and Pin 3. The same goes for any other numbers you're working with in your program. If you're counting the number of times a loop runs, 9 in the program means it has run ten times—counting 0, 1, 2, 3, 4, 5, 6, 7, 8, 9, for a total of ten numbers—rather than the nine you might expect. It can take a while to get used to thinking from zero, so don't worry if you forget!

There's more on using the BBC micro:bit's pins in Chapter 10, 'Building Circuits', Chapter 11, 'Extending the BBC Micro:bit', and Chapter 12, 'The Wearable BBC Micro:bit'.

When you've finished prodding the BBC micro:bit's Pin 0, remember to save your program with the Save button, move the `.py` file somewhere safe, and give it a memorable name like `touchinputs.py`.

Program 4: The Temperature Sensor

The BBC micro:bit has more inputs than simple buttons and touch-sensing pins. The most simple of these is the temperature sensor, which works exactly like a thermometer: read from the sensor and it returns a temperature, measured in degrees Celsius.

Although the temperature sensor is enough to give you a rough idea of how hot or cold your surroundings are, though, it's not a precision instrument. It's built into the BBC micro:bit's processor and was originally designed for monitoring the processor's temperature rather than environmental temperatures. If the BBC micro:bit were a desktop computer, that would be a problem: desktop computer processors run tens of degrees hotter than their surroundings even when idle, requiring large metal heatsinks and fans to keep them cool. The BBC micro:bit's processor, though, runs close enough to the temperature of its surroundings— known as the *ambient temperature*—that it is typically accurate to within a degree or two.

TIP

If you're looking to get the most accurate temperature reading possible, it's important to keep your program simple. If your program is working the BBC micro:bit's processor hard—doing lots of complicated sums, for instance—then the processor begins to warm up, throwing off your readings.

Start your program in the traditional way: save, move, and rename your existing program if you haven't already, and then press Ctrl+A on your keyboard followed by Delete to start with a fresh, blank program listing.

Type the import line to begin your program, followed by an infinite loop:

```
from microbit import *
while True:
```

Press Enter at the end of Line 2 to create a blank Line 3 with an automatic four-space indentation, which tells Python it will be part of the infinite loop you just created. Then type the following line of code:

```
display.scroll(str(temperature()))
```

This line represents a combination of three instructions (see Figure 7-12): `display.scroll`, which is used to scroll a string on the BBC micro:bit's display as it did in the 'Hello, World!' and Touch Inputs programs; `temperature()`; and `str` to convert the number—or integer—output by the `temperature` instruction into a string the `display.scroll` instruction can show on the BBC micro:bit's display. This highlights a key feature of programming in Python: rather than putting a number like `0` into the `display.scroll` instruction, or even a variable like `touches` earlier in this chapter, you can put a whole instruction in there and have it print out the instruction's output—in this case, the reading from the BBC micro:bit's internal temperature sensor.

FIGURE 7-12: Reading from the temperature sensor

You can now click the Download button to compile and download your program, flash it to the BBC micro:bit, and watch the temperature readings scroll across the screen.

Formatting the Output

The program is now, technically, complete, but it's not very easy to read. The numbers scroll across the screen continuously with no hint what they're measuring. To fix that, press Enter to add a new, blank Line 4 with an automatic four-space indentation and type in the following, making sure there's a space between the first single quote and the word `Celsius` (see Figure 7-13):

```
display.scroll(' Celsius')
```

```
1  from microbit import *
2  while True:
3      display.scroll(str(temperature()))
4      display.scroll(' Celsius')
```

FIGURE 7-13: Formatting the program's output

Click the Download button and flash the new, longer program to your BBC micro:bit. Watch the difference: rather than printing the temperature reading by itself, the BBC micro:bit begins by printing the temperature followed by the word `Celsius` with a space between the two—which is the reason for including a space when you typed `' Celsius'` in Line 4.

To see the temperature sensor in action, touch a metal surface to discharge any static electricity you may be holding (see Chapter 2, 'Getting Started with the BBC Micro:bit'). Then carefully and gently place the tip of your finger over the small black square labelled PROCESSOR on the back of the BBC micro:bit. After a few seconds, you should see the temperature reading start to rise. Remove your finger, and after a few more seconds it will drop back down again.

Always be careful when touching components on the back of the BBC micro:bit. If you have **WARNING** not properly grounded yourself by touching a metal surface first, you could fire a static shock into the sensitive circuitry and potentially damage your BBC micro:bit beyond repair.

When you've finished, remember to click the Save button, move the resulting file somewhere safe, and rename it to a descriptive name like `temperaturesensor.py`.

Program 5: The Compass Sensor

The BBC micro:bit's magnetic compass is one of the two sensors labelled on the back—the other being the accelerometer, detailed later in this chapter—and one of the most interesting. By reading the strength of the local magnetic field in three dimensions, it can work out the direction the BBC micro:bit is facing relative to magnetic north. This same technology is built into many smartphones and is used by mapping software to figure out which direction the user is facing.

The first step of this project should be obvious by now. Save your previous program if you haven't already, and then press Ctrl+A followed by Delete to begin with a blank, fresh program listing before typing in the import and infinite loop lines:

```
    from microbit import *
while True:
```

When you press Enter at the end of Line 2, Python automatically indents Line 3 by four spaces to indicate that it forms part of the infinite loop. Type the following:

```
    display.scroll('Heading %s' % compass.heading())
```

This line is a little dense, so it requires a bit of unpacking. The `display.scroll` section should be familiar. It tells Python to take a message, or string, and print it to the BBC micro:bit's display. In the Temperature Sensor program, you used this twice in quick succession to print the temperature reading followed by a helpful `Celsius` message. Here it's doing the same thing—formatting the message printed to the BBC micro:bit's display—in a single line: the message begins with the word `Heading` and ends with the actual sensor reading.

The secret to this trick is surprisingly simple. It all lies in the `%` symbol (see Figure 7-14). This tells Python that it should insert the contents of a variable or output of an instruction into another string. In this case, the first string is the word `Heading` followed by a space, and the second string is the output of the `compass.heading()` instruction. Notice how there are two `%` symbols in this line: the first, `%s`, is inside the single quotes as part of the string and tells Python to insert a string at that position—that's what the `s` of `%s` means, 'string'; the second is outside the single quotes and tells Python that what follows provides the string that should be placed where the `%s` is. What took you two lines in the Temperature Sensor program takes only one in this program.

```
1  from microbit import *
2  while True:
3      display.scroll('Heading %s' % compass.heading())
```

FIGURE 7-14: The completed program

Click the Download button to compile and download the program into a hex file, and then flash it onto the BBC micro:bit. Turn the micro:bot around, taking care not to dislodge the USB or battery cable, and you see the heading read by the compass change accordingly.

> **TIP** If this is your first time using the compass, or if you've moved from one environment to another since the last time you used it, you may be asked to *calibrate* it. If the message `Draw a circle` appears on the display when you run your program, simply do as asked: rotate and tilt the BBC micro:bit so that the dot in the centre of the display rolls around the edge of the display and draws a circle. When the circle is complete, the compass is calibrated and the program runs normally. The accuracy of the calibration is very dependent on environmental factors. If you're calibrating the compass in an area surrounded by metal or near magnetic fields—such as on a metal desk or next to a set of speakers—the compass will not calibrate correctly. For the highest accuracy, always try to calibrate the compass in an environment as close to that in which you'll be using it as possible.

One neat feature of using the `%` symbol to add to strings is that you're not limited to adding just one: `'1 %s %s' % ('2', '3')` results in `123`, whereas `'1 %s %s %s' % ('2', '3', '4')` gives you `1234` and so on. You can keep adding strings using the `%` symbol without ever needing to write a second line of code. Try it yourself by modifying the existing program to print out the word `degrees` after the compass reading.

As always, when you've finished, remember to click the Save button, move the resulting file somewhere safe, and rename it to a descriptive name like `compasssensor.py`.

Program 6: The Accelerometer Sensor

The second of the two sensors labelled on the rear of the BBC micro:bit, alongside the compass, is the *accelerometer*. Where the compass is designed to measure magnetic fields and thus determine which direction the BBC micro:bit is facing, the accelerometer measures relative acceleration in three planes: X, Y, and Z. As well as being able to return actual sensor values, it can be used to detect different types of motion known as *gestures*—by far the simplest way to interact with the sensor in your programs.

The accelerometer is a powerful tool. Besides being able to detect motion, it can be used to calculate the angle at which the BBC micro:bit is positioned by tracking the force of gravity pulling the BBC micro:bit toward the centre of the Earth—even when the BBC micro:bit is sat securely on a table or held in your hand. It's this feature of an accelerometer that a smartphone or tablet uses to detect when you've turned it from portrait to landscape orientation or vice versa, so that it can automatically rotate the display to match.

To begin, make sure you've saved, moved, and renamed your previous program. Then press Ctrl+A followed by Delete to start with a blank program listing.

> **TIP**
>
> A well-documented program is a good program. Although you won't be instructed to type comments during this chapter, as they're not required to make the program run, it's a good idea to get into the habit anyway. A part of the program that makes perfect sense today might seem confusing if you come back to the program a few months down the road and could be entirely impossible for someone else to understand without guidance. Comments don't slow a program down or increase the size of the compiled hex file; use them liberally and you'll save yourself—and anyone else who uses your program—a great deal of trouble down the road.
>
> To write a comment, simply type a hash symbol followed by the comment itself:
>
> # Like this
>
> A comment can be on a line by itself or at the end of an existing line of code. Anything to the right of the hash is ignored by the Python Editor and is not run, even if it's an otherwise valid line of code.

Start your program by typing the following four lines of code (see Figure 7-15), remembering that the Python Editor automatically handles the indentation for you:

```
from microbit import *
while True:
    if accelerometer.is_gesture("shake"):
        display.show(Image.SURPRISED)
```

```
1  from microbit import *
2  while True:
3      if accelerometer.is_gesture("shake"):
4          display.show(Image.SURPRISED)
```

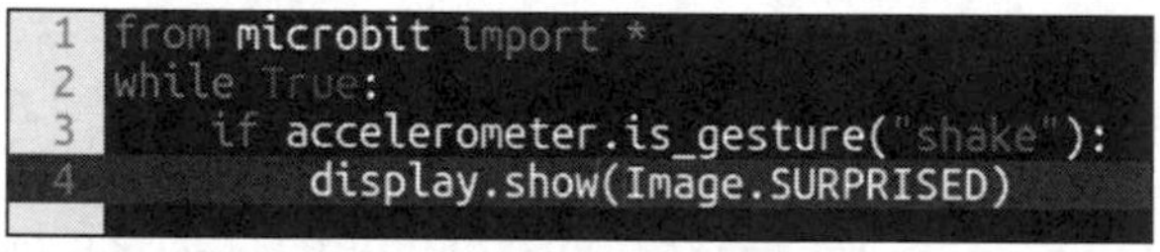

FIGURE 7-15: The is_gesture instruction

The program is now ready for testing, so click the Download button, drag the hex file to your BBC micro:bit, and wait for it to flash. When the flashing process is complete, hold your BBC micro:bit in your hand and carefully—making sure not to dislodge the micro-USB or battery cable—shake it to and fro. You see a surprised face appear on the BBC micro:bit's display. Well, wouldn't you be surprised?

The program works by using the accelerometer to look for one of a range of *gestures*, in this case the shake gesture. Shake is only one of a range of gestures the BBC micro:bit is programmed to recognise. The others follow:

- **8g**—The 8g gesture triggers only when the acceleration in any given direction exceeds eight times gravity, which is pretty fast.

- **6g**—The 6g gesture sits between the fast 8g and gentler 3g acceleration gestures.

- **3g**—The 3g gesture works in the same way as the 8g gesture, but it triggers at a more gentle rate of acceleration.

- **face up**—The face up gesture triggers when the BBC micro:bit's display is facing upward towards the user.

- **face down**—The opposite of face up, this gesture triggers only when the BBC micro:bit's display is facing downward.

- **freefall**—The free fall gesture triggers only when the BBC micro:bit has been dropped and is rushing toward the ground. This gesture is particularly useful when building flying robots or remote-controlled rockets.

- **shake**—Shake is activated by vigorously shaking the BBC micro:bit to and fro—but be careful you don't dislodge the micro-USB or battery cable or let go and send it flying across the room!

- **left**—This gesture is based on the angle of the BBC micro:bit and triggers when the BBC micro:bit is tilted towards the user's left.

- **right**—The opposite of left, the right gesture triggers when the BBC micro:bit is tilted to the user's right.

Delays

With a few changes, the program can become a lot more interactive. Move your cursor to the end of Line 4, if it isn't there already, and press Enter to add a new, blank Line 5. Then type the following, remembering that the Python Editor handles the indentation for you:

```
display.show(Image.ASLEEP)
```

If you click the Download button, flash your new program version, and then shake the BBC micro:bit, you briefly see the surprised image before the asleep image takes over. If you were to continuously shake the BBC micro:bit without stopping—the 'face' would change from the surprised image to the asleep image and back again over and over relatively quickly. That's not very realistic; it takes time to get back to sleep when you've been shaken awake!

To fix that, you're going to need to add a *delay* between the two `display.show` instructions. A delay does exactly what its name suggests: it pauses the program for a set length of time, delaying the execution of the next instruction. Move your cursor to the end of Line 4, press Enter to add a new, blank Line 5, and type the following (see Figure 7-16):

```
sleep(1000)
```

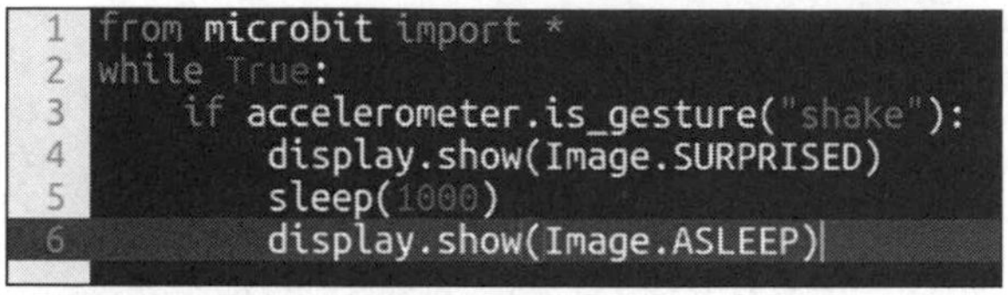

```
1  from microbit import *
2  while True:
3      if accelerometer.is_gesture("shake"):
4          display.show(Image.SURPRISED)
5          sleep(1000)
6          display.show(Image.ASLEEP)
```

FIGURE 7-16: The completed program

Click the Download button to download your program and then flash it onto your BBC micro:bit. This time, if you shake the BBC micro:bit, you see the surprised image longer before the asleep image takes over. Increase the number in the `sleep` instruction, and the surprised face stays on the display longer; decrease it and the asleep face appears sooner.

The number between the brackets in the `sleep` instruction is a measure of how long the program should delay in *milliseconds*, or thousandths of a second. A value of 1000, then, is equivalent to one second; 2000 is two seconds, and 500 is half a second. Because processors like the one powering the BBC micro:bit operate so much faster than humans, performing millions of operations every second, programs frequently need delays in them to allow the human operator to catch up.

As an experiment, see if you can extend the program still further by having the BBC micro:bit start on the asleep icon rather than starting entirely blank until the first shake. Hint: code within `if accelerometer.is_gesture("shake"):` only runs when the BBC micro:bit is shaken, so you need to put your code outside this statement.

When you've finished, remember to click the Save button, move the resulting `.py` file out of your Downloads folder somewhere safe, and rename it to a descriptive name like `accelerometersensor.py`.

Reading Raw Accelerometer Data

The gesture system is by far the easiest way of working with the BBC micro:bit's accelerometer, but it's not the only way. You can also read the raw data into your program as numbers to handle yourself. Create a fresh project by pressing Ctrl +A followed by Delete so you have a blank program listing.

Start by typing in the following three lines of code, beginning with the library import line without which the program simply won't run:

```
from microbit import *
while True:
    display.scroll('X:%s' % accelerometer.get_x())
```

This line is similar to the one you wrote in the Compass Sensor program. `display.scroll` tells the BBC micro:bit to print a string to the display, whereas the `%` symbol joins two strings: `"X:"` as a prefix followed by the output of the `accelerometer.get_x()` instruction.

Click the Download button and flash the resulting hex file to your BBC micro:bit. Try moving the BBC micro:bit around and watch the number change.

The X axis is only one of three axes that make up the accelerometer's three-dimensional measurements, though. To get at the others, click on the 3 at the start of Line 3 to highlight it and press Ctrl+C to copy it. Move your cursor to the end of Line 3, press Enter to add a new, blank Line 4, and then press Ctrl+V to paste the line of code you copied and create a new, blank Line 5. Press Ctrl+V again to create another copy, and then move your cursor to the right of the indentation in Line 4 and press Backspace to delete the extra four-space indentation that the Python Editor put in by mistake.

Your program is now set up to display the accelerometer data three times, but it's the same X axis data each time. To correct that, you need to make four changes. In Line 4 change `X:` to `Y:` and `get_x` to `get_y`, then in Line 5 change `X:` to `Z:` and `get_x` to `get_z` (see Figure 7-17).

```
1  from microbit import *
2  while True:
3      display.scroll(('X:%s' % accelerometer.get_x())
4      display.scroll(('Y:%s' % accelerometer.get_y())
5      display.scroll(('Z:%s' % accelerometer.get_z())
```

FIGURE 7-17: The completed program

Click the Download button and flash the hex file to your BBC micro:bit. When the program has flashed, the BBC micro:bit now reads out all three of the axes one after another: the X axis, followed by the Y axis, then the Z axis before returning the X axis again. Try moving the BBC micro:bit to see how its angle and position affect the numbers—measured in *microgravities*—shown on the display. With the BBC micro:bit facing you, tilting it left will reduce the number reported on the X axis, while tilting it right will increase it. Tilting it so the top is closer to you will reduce the number reported on the Y axis, while tilting it so the top is further away will increase it. Finally, lifting the BBC micro:bit up will increase the number reported on the Z axis, while lowering it down will reduce it (see Figure 7-18).

As always, when you've finished, make sure you click the Save button, move the `.py` file somewhere safe, and rename it to something like `accelerometersensordata.py` so you can find it again in the future.

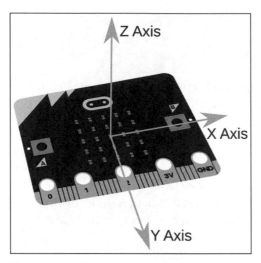

FIGURE 7-18: The accelerometer's three axes

Program 7: The Fruit Catcher Game

The programs in this chapter have been relatively simple, doing only one thing at a time, as a means of introducing some of the base concepts you need to know when programming the BBC micro:bit in Python. This final program, though, is considerably more complex: a game which has the player—in the form of a single glowing pixel at the bottom of the display—attempt to catch ever-faster falling fruit—another pixel, starting at the top of the display and falling downwards—before it hits the bottom of the display.

The Fruit Catcher game is relatively simple, but it's still quite a hefty program. If you've been working through this chapter nonstop, now might be a good time to take a break away from the screen so you can come back refreshed.

As ever, the game starts the same way as any program: saving, moving, and renaming any existing program you have loaded in the Python Editor and then pressing Ctrl+A followed by Delete to start with a fresh program listing.

> **TIP**   If you'd prefer to type the program out in one go and then read about what each line does afterward, you can find a full copy of the program code in Appendix C, 'Python Recipes'.

The Setup

Start your program by importing the `microbit` library, but add a second line below it by typing the following:

```
from microbit import *
import random
```

The first line should be familiar by now, but the second is new: `import random` does exactly what it says, importing the library called `random` into the program. You use this later to inject randomness into the game to make it more interesting, but the key thing to note here is that you can have both the `microbit` and `random` libraries added at the same time. If you find you need a feature that isn't in the `microbit` library, as with this game, simply add another `import` line at the top of the program to import the library with the function or functions you need.

Next, you're going to need to initialise some of the variables the game needs: a variable for adjusting the speed of the game called `delay`, a `delayCounter` variable that you will use to adjust the speed of the game during play, and a variable to hold the position of the player. Type the following lines directly beneath Line 2:

```
delay = 10
delayCounter = 0
playerPosition = [2, 4]
```

The first line sets the game's timing delay, which is multiplied by 100 milliseconds later in the program giving an initial delay of one second. This line demonstrates an important feature of variable initialisation: you can initialise a variable with whatever data you require, not just zero as in previous programs. The delay initialised in this line controls how fast the game is when it starts. If you'd like a challenge, you can make it lower, or to make the game easier, make it higher.

The second line (of the snippet you just typed) initialises the `playerPosition` variable, which the game uses to control where the player—represented by a dot—is drawn on the screen. Notice how this variable has two numbers, enclosed in square brackets and separated by a comma. This represents a *list*, which is effectively a variable which can hold more than one piece of data (or *datum*)—in this case, two numbers that would otherwise have had a variable each.

The numbers in the `playerPosition = [2, 4]` variable are important. They control exactly where the player, known as a *sprite*, appears on the BBC micro:bit's display. Every pixel on the display has a location on the horizontal X axis and the vertical Y axis. Figure 7-19 demonstrates this, giving the X and Y coordinates as "X,Y" for each of the 25 onscreen pixels.

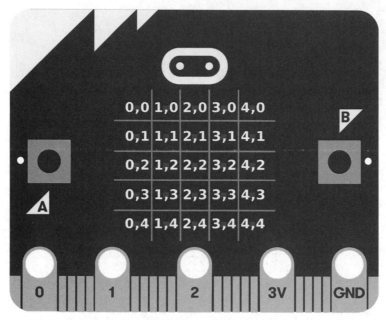

FIGURE 7-19: The BBC micro:bit Display Coordinates

By setting the `playerPosition` instruction to position the player sprite at position 2 on the X axis and 4 on the Y axis—coordinate 2,4 in Figure 7-19—the sprite is created in the middle column of the bottom row of the display.

Next, you need to initialise the game's score at zero. You could, if you wanted, cheat at this point and have the game start on a score of 10, or 100, or any other number you like, but where's the fun in that? Move your cursor to the end of Line 5, if it isn't there already, press Enter, and type the following line of code:

```
score = 0
```

With Line 5 now in place and a starting score initialised, the setup portion of your game is complete (see Figure 7-20).

```
1  from microbit import *
2  import random
3  delay = 10
4  delayCounter = 0
5  playerPosition = [2, 4]
6  score = 0
```

FIGURE 7-20: The finished setup portion

The Main Program Loop

The next stage of the process is to make the game. Move your cursor to the end of Line 6, if it isn't there already, and press Enter to insert a blank Line 7 followed by Enter again to put your cursor on a blank Line 8. The game needs to take place inside a loop, so create one and insert the first instruction by typing this:

```
while True:
    fruitPosition = [random.randrange(0,4), 0]
```

> **TIP**
> The blank Line 7 is in your program for a reason: adding spaces between sections of code makes it easier to see where one section ends and another begins. Blank lines, like comments, don't slow your program down or make the finished program any bigger. They make it considerably easier for someone else—or yourself—to understand when you open the project in a few months' time and try to remember which section is which!

If the fruit appeared in the same place every time you played the game, it wouldn't be much of a game. Where the player sprite's starting position was set back in Line 5, Line 9 uses the `randrange` instruction from the `random` library you imported on Line 2 to find a random starting position every time a new fruit sprite is created.

The `random.randrange` instruction does exactly what it says: picks a random number starting at the first number between the brackets and ending at the second number between the brackets, which in this case is zero and four. This gives 0, 1, 2, 3, and 4—five numbers in total—as the possible output of the `random.randrange` instruction. That range of five numbers maps perfectly to the X or Y axis of the display. That's how it's used here: when the game creates the fruit sprite, it positions it at the top (Y: 0) of the display but chooses a random location in that top row (X: 0 to 4, depending on what the random number generator picks).

At this point, the program should look like Figure 7-21.

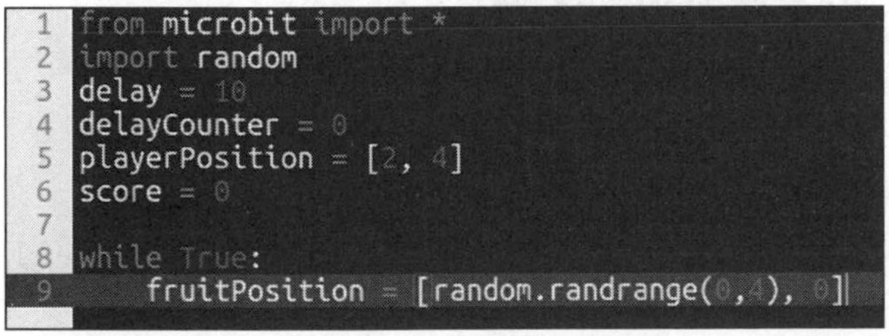

```
1  from microbit import *
2  import random
3  delay = 10
4  delayCounter = 0
5  playerPosition = [2, 4]
6  score = 0
7
8  while True:
9      fruitPosition = [random.randrange(0,4), 0]
```

FIGURE 7-21: The beginnings of the main program

Conditional Loops

So far, you've only worked with a single type of loop: the infinite loop, which runs forever. A more powerful type of loop is the *conditional loop*, which tests for a particular condition—such as whether a variable is equal to another variable—and runs only when the condition is true. A conditional loop works a lot like an if statement, except that whereas an if statement comes to the end of its indented lines of code and finishes, a conditional loop goes back to the beginning and tests its condition again to see if it should repeat.

Move your cursor to the end of Line 9, if it isn't there already, and then press Enter and type the following:

```
while fruitPosition[1] < 4:
```

Press Enter when you've typed the line above, and it leaves you on Line 11 with a new level of indentation: eight spaces instead of four, which indicates that the block of code you're typing next sits under the new loop you just created.

This line opens a while loop, a conditional loop that only runs when the condition it is testing is true, and retrieves the fruitPosition[1] variable—the second entry, counting from zero, of the fruitPosition list created a line earlier, which holds the location on the display's Y axis of the fruit sprite. It then takes this value, between zero for the top row of the display and four for the bottom row of the display, and checks to see if it is less than (<) four (see Figure 7-22). If, and only if, it is—which is to say that the fruit sprite hasn't yet reached the bottom of the screen—do the lines of code within the loop run.

```
 1  from microbit import *
 2  import random
 3  delay = 10
 4  delayCounter = 0
 5  playerPosition = [ , ]
 6  score = 0
 7
 8  while True:
 9      fruitPosition = [random.randrange( , ), 0]
10      while fruitPosition[1] <= 4:|
```

FIGURE 7-22: The conditional loop

At the moment, though, the loop doesn't actually do anything. With your cursor on Line 11, after the eight-space indentation, type the following:

```
while delayCounter < delay:
```

This line creates another conditional loop, this time testing to see if the value of the `delayCounter` variable is less than the value of the `delay` variable. This is used to control how fast the game runs. Each time this loop runs, the `delayCounter` variable is incremented, or increased by one; when it reaches the value stored in the `delay` variable, set to 10 on Line 3, the loop finishes and the rest of the program runs.

Conditional Statements

To catch the fruit, the player needs to be able to move left and right along the bottom of the BBC micro:bit's display. Fortunately, there's Button A to the left of the display and Button B to the right—just what you need for controlling the player sprite. Reading these buttons and controlling the player sprite is effectively a reprise of the Button Inputs program you wrote earlier in the chapter, but with a little extra logic to help things along.

With your cursor at the end of Line 11, press Enter to create a new Line 12 with a twelve-space indentation before typing the following:

```
            if button_a.was_pressed() and↩
(playerPosition[0] > 0):
```

When a line of code would extend past the border of the page, a ↩ symbol is printed. When **TIP** you see this symbol, continue to type the code without pressing the Enter or Return keys. If you're not sure how a line of code should be entered, visit the website at www.wiley.com/go/bbcmicrobituserguideto download plain-text versions of each program; these can then be used for reference or even simply copy and pasted directly into the editors.

The conditional on this line exists to stop the player sprite leaving the left edge of the display. It first checks to see if Button A, on the left side of the display, has been pressed. Then it checks the player's current position on the X (horizontal) axis: if it's greater than 0, representing the left edge of the display, the player should be allowed to move further left; if not, the player is already at the leftmost edge and should not be allowed to move farther left.

You may notice that this conditional uses `button_a.was_pressed()` rather than `button_a.is_pressed()` as was used for the Button Inputs program. That makes an important distinction: `button_a.is_pressed()` checks that the button is being pressed only when the conditional is actually being tested; `button_a.was_pressed()` checks whether the button was pressed at any time prior to the last time a `button_a.was_pressed()` check was made. The reason for using this is explained later in this chapter.

To actually make the player sprite move, make sure your cursor is still on Line 12, press Enter to create a Line 13 with a sixteen-space indentation, and type the following:

```
                playerPosition[0] -= 1
```

Now when the player presses Button A but isn't already at the left edge of the screen, the player's position on the X axis—stored in the first entry in the `playerPosition` list, while the second stores the position on the Y axis—is decreased by one (or *decremented*), moving the sprite one pixel to the left.

Create another conditional watching for the player pressing Button B to indicate a desire to move right by typing the following lines, making sure to delete four spaces from the automatically created 16-space indent on Line 14 by pressing Backspace once before typing this:

```
        if button_b.was_pressed() and↵
 (playerPosition[0] < 4):
                playerPosition[0] += 1
```

Remember to enter all lines with a ↵ as one line. Now the game has full control: the first statement watches for Button A being pressed, checks that the player has room to move to the left, and then moves the player sprite leftward one column. The second statement watches for Button B being pressed, checks that the player has room to move to the right, and then moves the player sprite rightward one column.

Drawing the Sprites

Previously, your interactions with the display were limited to using predrawn images or making messages scroll across. Making a game requires that you can build an image on-the-fly, though, which is achieved using the `display.set_pixel` instruction. With your cursor at the end of Line 15, press Enter to create a new Line 16, and then press Backspace until you have erased 4 of the 16 spaces that make up the indentation, leaving 12 spaces remaining. Then type the following:

```
display.clear()
            display.set_pixel(fruitPosition[0], ↵
 fruitPosition[1], 9)
            display.set_pixel(playerPosition[0], ↵
 playerPosition[1], 9)
```

The first line clears the BBC micro:bit's display, ensuring that any previously drawn sprites don't linger. The second takes the X and Y coordinates of the fruit sprite as stored in the fruit Position array to pick a location on the screen—as shown in Figure 7-19 earlier in this chapter—then sets its brightness to 9, the maximum possible value. The third line, finally, does the same for the X and Y coordinates of the player sprite as stored in the `playerPosition` array, again setting that pixel of the display to maximum brightness.

At the rate the BBC micro:bit processor runs through Python instructions, a game where the fruit falls as quickly as possible would be many millions of times faster than a human could play; you'd see nothing but the 'game over' screen. To fix that, with your cursor at the end of Line 18, press Enter, and type the following:

```
delayCounter += 1
sleep(100)
```

This increments the `delayCounter` variable, which controls how many times this conditional loop will run, before pausing the program for 100 milliseconds. Remember that the conditional loop, opened on Line 11, compares the `delayCounter` variable to the `delay` variable initialised at 10 on Line 3, meaning that the loop runs a total of 11 times (with the `delayCounter` variable counting 0, 1, 2, 3, 4, 5, 6, 7, 8, 9, and finally 10), multiplying the 100 millisecond delay by 11 for a total delay of 1.1 seconds.

This may seem like a long-winded way of slowing down the program when you could just write `sleep(1100)` and achieve the same thing, but there's a reason for doing it this way: when the Python program is paused by a sleep(1100) instruction, it's not checking to see if the buttons are being pressed. If it's not checking to see if the buttons are being pressed, the player can't move. Without this loop, which keeps checking to see if the player wants to move even while waiting to make the fruit fall down the screen, the game would be unresponsive and not much fun to play.

This is also why the game uses `button_a.was_pressed()` to check when the button was pressed. During a delay, the Python program isn't doing anything else—including checking for button presses. By using `button_a.was_pressed()`, the player sprite is moved even if the player pressed the button during the `sleep(100)` part of the program—which is more than likely, because due to the speed of its processor the game spends most of its time in a delay rather than running through its program code.

Finishing the Program

Press Enter at the end of Line 20, and then press Backspace once to delete four of the spaces the Python Editor has automatically added for indentation to make the next lines you type part of the loop opened on Line 10 rather than Line 11. This loop controls the falling of the fruit, delayed by the loop opened on Line 11 to compensate for the speed of the BBC micro:bit's processor. Type the following two lines:

```
delayCounter = 0
fruitPosition[1] += 1
```

The first line resets the `delayCounter` variable back to zero, so the next time the game enters the delay loop opened in Line 11, it runs through from a starting count of zero rather than immediately exiting with the already-stored `delayCounter` value of 10.

The second line is responsible for making the fruit drop: the Y coordinate, stored in the second entry of the `fruitPosition` array, is incremented, moving it one pixel closer to the bottom of the screen. This represents the end of the main game loop. When this line is run, Python goes back to the start of the loop on Line 10 and checks to see if the position of the fruit has reached the bottom row of the display. If it hasn't, the loop continues and everything starts again; if it has, the loop ends and the program moves on to Line 23—a line you now need to write.

With your cursor at the end of Line 22, press Enter followed by the Backspace key once to delete four of the spaces the Python Editor has inserted as indentation to indicate that the line you are writing is not part of the loop opened on Line 11. Then type the following:

```
if fruitPosition[0] == playerPosition[0]:
```

This line checks whether the position of the fruit sprite on the X axis, stored in the first entry of the `fruitPosition` array, is equal to the position of the player sprite on the same axis, stored in the first entry of the `playerPosition` array. If the player sprite and the fruit sprite are in the same location, the player has caught the fruit; if not, the player has missed the fruit.

> **TIP** When looking for two things that are equal in a conditional, you must always use == rather than =. In Python, a single = is used to set the value of a variable; == is used to compare two variables, or other items, to see if they are equal.

If the player has caught the fruit, the player's score needs to be increased. With your cursor at the end of Line 23, press Enter and keep the Python Editor's automatic indentation. Then type the following:

```
score += 1
delay -= (delay / 10)
```

It's the second line of code that is responsible for speeding up the game after every successful catch of the fruit. The first part of the code tells Python to take away from the `delay` variable whatever is found in the second part; the second part itself tells Python to calculate the current value of the `delay` variable divided by 10. The result: the `delay` variable shrinks by 10 percent each time the fruit is caught, making the game 10 percent faster each round.

This conditional handles the case of the player successfully catching the fruit, but the game needs to be able to run a different instruction if the player misses. That's handled by adding an `else` case to the conditional. Press the Enter key to insert a new, blank Line 26, press Backspace to delete four spaces from the indentation, and then type the following two lines:

```
    else:
        display.scroll(('GAME OVER    SCORE %s' %↵
 score), loop=True)
```

Read as a whole, these instructions form the following sentence in plain English: 'if the location of the fruit sprite is equal to the location of the player sprite, add one to the score and decrease the delay by 10 percent; if it is not equal, end the game and display a game over screen'.

Code in the `if` portion of the conditional only runs if the conditional is true; code in the `else` portion of the conditional only runs if the conditional is false. No matter what happens, only one of the two blocks of code will ever run; under no circumstances will the program ever run both the `if` and the `else` statements on a single run-through of the loop.

Although you've used the `display.scroll` instruction before, the `loop=True` section is new. This tells `display.scroll` that it should repeat the scrolling message once it has reached the end, which completes the game. Without `loop=True`, Python would display the message and then start the game again. With it, you have to restart the game yourself by pressing the BBC micro:bit's Reset button.

Before trying out your new game, check that the program has been entered correctly by comparing it to the version found at the back of this book in Appendix C, or Figure 7-23, and making any changes necessary. Once you're sure the program is correct, you can click the Download button, flash the hex file onto your BBC micro:bit, and start to play.

Move the player sprite, at the bottom of the screen, with Button A and Button B, and try to catch the fruit sprite as it falls from the top to the bottom. Miss the fruit and it's game over; you'll see a GAME OVER message and your score. To restart the game, just hit the Reset button again.

To improve the game—and your programming skills—try making some modifications. Instead of controlling the player sprite with the buttons, how about using the accelerometer to watch for tilting and give the game motion control? Try adding multiple fruit which all fall down the screen at once.

```
1  from microbit import *
2  import random
3  delay = 10
4  delayCounter = 0
5  playerPosition = [2, 4]
6  score = 0
7
8  while True:
9      fruitPosition = [random.randrange(0,4), 0]
10     while fruitPosition[1] <= 4:
11         while delayCounter < delay:
12             if button_a.was_pressed() and (playerPosition[0] > 0):
13                 playerPosition[0] -= 1
14             if button_b.was_pressed() and (playerPosition[0] < 4):
15                 playerPosition[0] += 1
16             display.clear()
17             display.set_pixel(fruitPosition[0], fruitPosition[1], 9)
18             display.set_pixel(playerPosition[0], playerPosition[1], 9)
19             delayCounter += 1
20             sleep(100)
21         delayCounter = 0
22         fruitPosition[1] += 1
23     if fruitPosition[0] == playerPosition[0]:
24         score += 1
25         delay -= (delay / 10)
26     else:
27         display.scroll(('GAME OVER    SCORE %s') % score, loop=True)
```

FIGURE 7-23: The finished game

Further Steps

If you skipped straight to this chapter, see how the same programs are implemented in a visual development environment in Chapter 5 or in the JavaScript Language in Chapter 6. If you'd prefer to stick with Python, you'll find a range of sample programs, exercises, and projects on the Micro:bit Educational Foundation website at microbit.org.

Part III

Advanced BBC micro:bit Projects

Chapter 8

The Wireless BBC micro:bit

ONE OF THE most powerful features of the BBC micro:bit is its built-in radio module. Built into its processor, a small black chip on the back of the board, the radio might not look like much, but it's extremely powerful. Using the radio, you can link two BBC micro:bits for wire-free communication or even have groups of dozens of BBC micro:bits all sharing data.

Best of all, the BBC micro:bit comes with its radio ready to run straight out of the box. There's no need to connect an external antenna, and the radio frequencies used are licensed for free use around the world. There's nothing to stop you from making your next project a wireless one.

The BBC micro:bit Radio

The BBC micro:bit's radio module is built into the processor, a small black chip found on the back of the board to the left side, and marked with the 'PROCESSOR' label on the silkscreen layer (see Figure 8-1). The module itself isn't labelled, as it is integrated directly into the processor, but its antenna is with the marking 'BLE ANTENNA' pointing to the upper-left corner of the board where specially-shaped circuit board traces act to receive and send radio signals.

FIGURE 8-1: The BBC micro:bit processor with radio module

Although there's only one physical chip, there are actually two radio functions provided. The first function provides what is known as *peer-to-peer connectivity*, sending signals which can be received by any nearby BBC micro:bit. The second function uses Bluetooth Low Energy (BLE), a variant of the Bluetooth short-range wireless standard found in all smartphones and most tablets but modified to ensure the bare minimum power usage.

The peer-to-peer radio function is the one to use if you want to create projects where BBC micro:bit modules communicate with each other. Whether you're building a BBC micro:bit-powered doorbell, a multiplayer game, or a pager alert system, the peer-to-peer radio allows you to get started quickly and easily.

Program 1: One-to-One Communication

The simplest use of the BBC micro:bit radio module is for one-to-one communication: sending a message from one BBC micro:bit and receiving it on another. This is known as peer-to-peer communication, as there is no central server. The first BBC micro:bit talks directly to the second BBC micro:bit, and the second is free to talk to the first. Setting up this kind of wireless network takes just a few lines of code. To follow along with this example, you need two BBC micro:bits with micro-USB cables or battery packs for power.

TIP The programs in this chapter are written using the JavaScript language. Before attempting to use any of them, you should have completed the exercises in Chapter 6, 'JavaScript', to familiarise yourself with the JavaScript Editor and the language itself. If you'd prefer to use JavaScript Blocks or Python, implementations of each program in these languages can be found in Appendix A, 'JavaScript Blocks Recipes', and Appendix C, 'Python Recipes', respectively. You can also find full copies of each JavaScript implementation from this chapter in Appendix B, 'JavaScript Recipes'.

To configure a BBC micro:bit for radio use, you need only one line of JavaScript. Go to `makecode.microbit.org` in your web browser, click the Projects menu and New to start a new project, click the Editor toggle to switch to JavaScript Mode, rename the program 'Radio Experiments', and delete the lines the JavaScript editor has already filled in before typing the following (see Figure 8-2):

```
radio.setGroup(1)
```

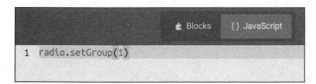

FIGURE 8-2: Setting up the peer-to-peer radio module

That simple line assigns the BBC micro:bit a *radio group*. You learn more about radio groups later in this chapter, but for now it's enough to know that both the BBC micro:bits must be set to the same radio group for the program to work. Although best practice is to set the radio group manually, as in this example, it's possible to write a program which uses the radio but doesn't include a `radio.setGroup` instruction. In these cases, the BBC micro:bit will be automatically assigned to a default group as soon as the radio is used by your program. This default group is controlled by your program itself: if you load the same program onto two BBC micro:bits, they will both be in the same default group; make any changes to the program before loading it onto the second BBC micro:bit, though, and the two will be in different groups and unable to communicate with each other. Next, set up an event (see Chapter 6) to be triggered when Button A is pressed:

```
input.onButtonPressed(Button.A, () => {
    radio.sendString("Hello from A!")
})
```

The `radio.sendString` instruction does exactly what its name suggests: activates the radio and sends, or *transmits*, a message in the form of a string of text, in this case the message `"Hello from A!"` If you pause after typing **radio.** and view the JavaScript Editor's pop-up, you see the other instructions for controlling the radio, one of which you need to use next to create an event for handling receiving, rather than transmitting, messages. Type the following:

```
radio.onDataPacketReceived(({receivedString}) => {
    basic.showString(receivedString)
})
```

This event is triggered when the radio receives a message from another BBC micro:bit in the same radio group. The `basic.showString` instruction is then used to print the message to the BBC micro:bit's display. Your first radio program is now finished (see Figure 8-3).

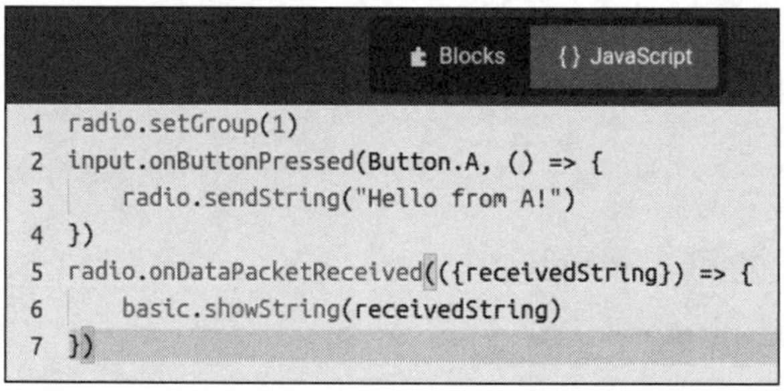

```
1  radio.setGroup(1)
2  input.onButtonPressed(Button.A, () => {
3      radio.sendString("Hello from A!")
4  })
5  radio.onDataPacketReceived(({receivedString}) => {
6      basic.showString(receivedString)
7  })
```

FIGURE 8-3: The completed one-to-one program

Click the Download button to compile the program. Then flash the hex file to your first BBC micro:bit, which we'll call BBC micro:bit A (see Chapter 3, 'Programming the BBC micro:bit', for details). When the flashing process is finished, press Button A. The radio transmits its message, but you've no way to receive it. The program is written so that it is either transmitting its message or listening for a message, and when it's transmitting it isn't listening—and thus can't receive the message it is sending. For that, you need a second BBC micro:bit.

Go back to your web browser with the JavaScript Editor loaded, and then modify Line 3 as follows:

```
radio.sendString("Hello from B!")
```

The rest of the program should remain unchanged (see Figure 8-4). Click the Download button again to compile the program, and then flash the new hex file to your second BBC micro:bit. With both BBC micro:bits switched on, press Button A on BBC micro:bit A; you see the message `Hello from A!` scroll across BBC micro:bit B's display. Press Button A on BBC micro:bit B, and the opposite happens with the message `Hello from B!` scrolling across BBC micro:bit A's display.

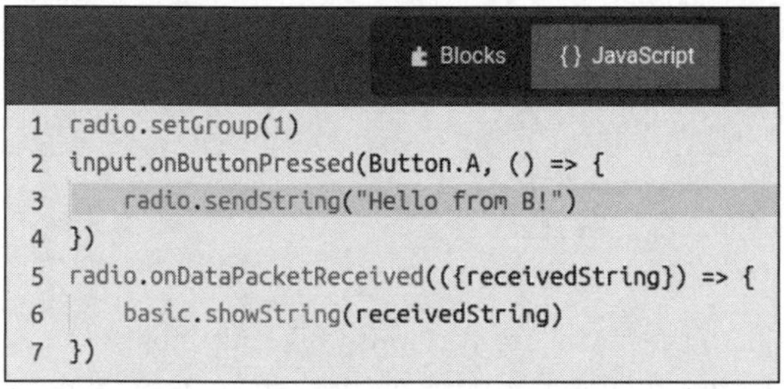

```
          Blocks    {} JavaScript
1  radio.setGroup(1)
2  input.onButtonPressed(Button.A, () => {
3      radio.sendString("Hello from B!")
4  })
5  radio.onDataPacketReceived(({receivedString}) => {
6      basic.showString(receivedString)
7  })
```

FIGURE 8-4: Modifying the program for a second BBC micro:bit

This program may seem simple, but it hides real practicality. With one BBC micro:bit by the door and the other by your desk, you can tweak the messages displayed to turn the program into a wireless two-way doorbell. When there's someone at the door, a DING-DONG message—or even an actual bell sound, if you add extra hardware to the BBC micro:bit's input output pins—scrolls across the BBC micro:bit on your desk. Confirm receipt of the message with a button push, and the person at the door sees On my way! scroll across the display.

The BBC micro:bit's radio module is not limited to communication between just two BBC micro:bits, either, as you'll discover in the next program.

Program 2: One-to-Many Communication

To follow along with this example, you need a third BBC micro:bit. Go back to the JavaScript Editor, and then edit Line 3 again so the message from the third BBC micro:bit is unique. When you're finished, the program should look like this (see Figure 8-5):

```
radio.setGroup(1)
input.onButtonPressed(Button.A, () => {
    radio.sendString("Hello from C!")
})
radio.onDataPacketReceived(({receivedString}) => {
    basic.showString(receivedString)
})
```

```
                                    ⚫ Blocks    {} JavaScript
1  radio.setGroup(1)
2  input.onButtonPressed(Button.A, () => {
3      radio.sendString("Hello from C!")
4  })
5  radio.onDataPacketReceived(({receivedString}) => {
6      basic.showString(receivedString)
7  })
```

FIGURE 8-5: Modifying the program for a third BBC micro:bit

Click the Download button, and then flash the hex file to your third BBC micro:bit. With all three BBC micro:bits switched on, start by pressing Button A on BBC micro:bit A. You see the message Hello from A! scroll across the display of BBC micro:bit B and BBC micro:bit C. Press Button A on BBC micro:bit B, and the message Hello from B! scrolls across the display of BBC micro:bit A and BBC micro:bit C. Finally, press Button A on BBC micro:bit C; the message Hello from C! scrolls across the display of BBC micro:bit A and BBC micro:bit B.

Although you've built a *radio network* with three BBC micro:bits in this example, you can make the network as large as hundreds of BBC micro:bits. Using the example program, whatever one BBC micro:bit transmits is received and displayed by every other BBC micro:bit in the network—so long as they are all in radio range of each other, of course.

Because the network is peer-to-peer, with the BBC micro:bits communicating directly with each other, it is described as *decentralised*. If you unplug BBC micro:bit A, BBC micro:bits B and C can still communicate. Plug BBC micro:bit A back in and unplug BBC micro:bit B, and BBC micro:bits A and C are able to communicate; plug BBC micro:bit B back in and unplug BBC micro:bit C, and BBC micro:bits A and B can still talk just fine. Technically, the network still operates even when there's only one BBC micro:bit, but transmitting a radio message when there's nothing around to receive it is as useful as talking to an empty room.

Figure 8-6 shows a sample network with four BBC micro:bits. The arrows represent each BBC micro:bit's lines of communication. Notice how any one BBC micro:bit can talk to any other BBC micro:bit, even if you take the other BBC micro:bits away. When the BBC micro:bit at the bottom of the figure transmits a message, it is visualised by the arrow: the message is received by all three of the other BBC micro:bits, without having to go through a central hub, router, or server.

To extend the doorbell practical implementation example from Program 1, multiple BBC micro:bits could be connected this way to form a notification system for a multiroom building. Press the button on the BBC micro:bit nearest you, and every other BBC micro:bit displays the message. Alternatively, by removing the ability to receive messages from all but one BBC micro:bit, you can turn it into a one-way notification system: someone in one of many rooms presses the button on her nearest BBC micro:bit, and a message detailing which room is calling for assistance appears on a 'master' BBC micro:bit.

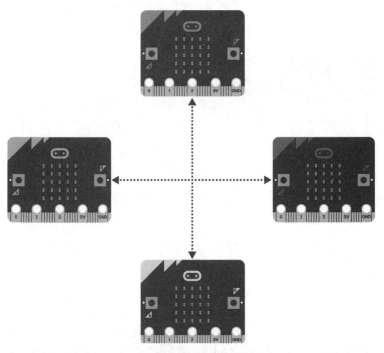

FIGURE 8-6: A decentralised network with four BBC micro:bits

Program 3: Radio Groups

Sometimes you don't want all the BBC micro:bits in a given area to be talking to each other. Perhaps you're in a classroom environment with children all trying to build radio networks which are interfering with each other. Perhaps your neighbour is using BBC micro:bits as well and you keep treading on each other's networks. Or perhaps you're building a project which uses many BBC micro:bits and you need a way to keep them organised.

Thankfully, the BBC micro:bit is ready for this: it supports radio groups, which allow you to create up to 256 individual radio networks with one or more BBC micro:bits per network. These groups don't affect the radio module directly—they don't change the *frequency* at which the radio transmits or receives, for example—but instead instruct the BBC micro:bit to ignore any transmissions which do not match the group number set in the program. All nearby BBC micro:bits will still receive any and all radio messages sent by any other nearby

BBC micro:bit, but if they are in different radio groups, the message is discarded—effectively like it was never received in the first place.

In previous programs, you have set the radio group to 1 (see Figure 8-7), one of the 256 possible radio groups ranging from group 0 to group 255. If you found that others using BBC micro:bits in your vicinity were intruding on your network, you could change this to any other number and filter out their messages.

FIGURE 8-7: Setting the radio group

It's also possible to modify your radio group on-the-fly—much like you might adjust the channel of a walkie-talkie radio if you start getting other people interrupting your conversations. Go back to the JavaScript Editor, and add the following lines to the end of the program:

```
input.onButtonPressed(Button.B, () => {
    radio.setGroup(2)
    basic.showString("Switched to Group 2")
})
```

Remember to change the `radio.sendString` message on Line 3 to say A (see Figure 8-8); then click Download and flash the resulting hex file to the first BBC micro:bit. Change Line 3 to read B, and then click Download again and flash the hex to the second BBC micro:bit. Finally, change Line 3 to read C. Then click Download one last time and flash the hex file to the third BBC micro:bit.

```
 1  radio.setGroup(1)
 2  input.onButtonPressed(Button.A, () => {
 3      radio.sendString("Hello from A!")
 4  })
 5  radio.onDataPacketReceived(({receivedString}) => {
 6      basic.showString(receivedString)
 7  })
 8  input.onButtonPressed(Button.B, () => {
 9      radio.setGroup(2)
10      basic.showString("Switching to Group 2")
11  })
```

FIGURE 8-8: The finished program, ready for BBC micro:bit A

Testing the Group Feature

Connect all three BBC micro:bits to power, either via the micro-USB port or the battery connector. Start by pressing Button A on each in turn. You see the same effect as in Program 2, with the message from the transmitting BBC micro:bit being received and displayed on both of the other BBC micro:bits. Press Button A on BBC micro:bit C, and you see the same thing happen.

Now press Button B on BBC micro:bit C, wait for the message confirming that the radio group has been switched, and then press Button A on BBC micro:bit A again. This time you see something different: the message `Hello from A!` scrolls across the display of BBC micro:bit B, but not BBC micro:bit C. This is because the three BBC micro:bits are now split across two different groups: BBC micro:bits A and B are in group one, while BBC micro:bit C is on its own in group two. You can confirm this in the other direction: press Button A on BBC micro:bit C again, and no message appears because BBC micro:bit C is in a group all by itself with nothing to receive the message it has transmitted. Remember that changing groups doesn't alter the radio frequency; all the BBC micro:bits are still technically receiving the radio signals sent by BBC micro:bit C, but because it is in a different radio group the messages are ignored.

Figure 8-9 demonstrates what's happening. What was a three-BBC micro:bit network is now two networks: one with two BBC micro:bits in it and the other with one BBC micro:bit in it. The lines of communication between BBC micro:bit C and the other two have been cut. The arrow in Figure 8-9 indicates possible directions of communication: BBC micro:bit A and BBC micro:bit B are free to act on each other's signals, but anything sent to or from BBC micro:bit C will be ignored.

Press Button B on BBC micro:bit A now, followed by Button A when the confirmation message has stopped scrolling. Now you see the message `Hello from A!` scrolling across BBC micro:bit C instead of BBC micro:bit B, because both BBC micro:bit A and BBC micro:bit C are now in group two (see Figure 8-10). As before, the arrow indicates possible directions of communication: BBC micro:bit A and BBC micro:bit C are now free to communicate, as they are in the same radio group, whereas BBC micro:bit B—left behind as the only member of the original radio group—is cut off.

BBC micro:bit B, meanwhile, is still in the original group one, now entirely on its own. Press Button B on BBC micro:bit B, and it joins the others in group two: now, pressing Button A on any of the BBC micro:bits shows the message across both of the other two BBC micro:bits (see Figure 8-11). As all three BBC micro:bits are now in the same group, any BBC micro:bit is free to communicate with any other BBC micro:bit. A signal sent from BBC micro:bit A will be received and acted upon by BBC micro:bits B and C.

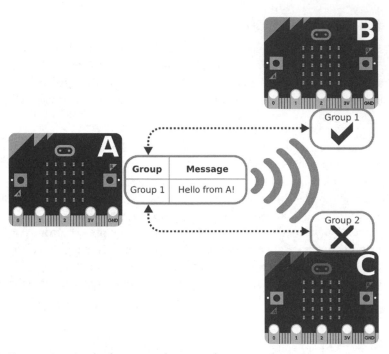

FIGURE 8-9: A split three micro-bit network

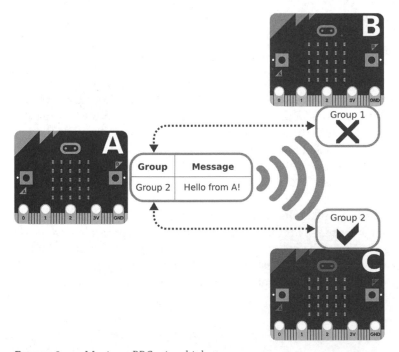

FIGURE 8-10: Moving a BBC micro:bit between groups

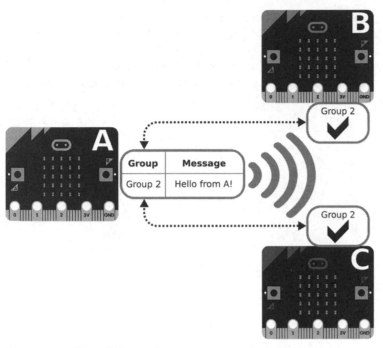

FIGURE 8-11: Three BBC micro:bits communicating in radio group two

> If you're having difficulty getting the concept of radio groups clear in your head, imagine each group is a room. Two people talking in the living room can hear each other fine, but the person in the bedroom has no idea what they are saying. If one person moves from the living room to the bedroom, the two people now in the bedroom can talk but the person left in the living room can't hear them. Only people in the same room can hold a conversation, just as only BBC micro:bits in the same radio group can form a radio network.
>
> **TIP**

With 256 groups at your disposal, it's possible to set up extremely dense networks with hundreds of BBC micro:bits and never run into problems with messages going somewhere they shouldn't.

Using the BBC micro:bit with a Smartphone or Tablet

The BBC micro:bit's Bluetooth Low Energy radio can also be used to communicate with a smartphone or tablet running the Google Android or Apple iOS operating system. By linking the BBC micro:bit to a compatible smartphone or tablet, you can interact with the it—including reading from its sensors or transmitting messages to be displayed—and even flash new programs onto it without having to connect a micro-USB cable.

To link a smartphone or tablet to a BBC micro:bit, you need to install an app on your device. For the latest instructions on how to install the app and pair the BBC micro:bit to your device, visit `microbit.org/guide/mobile`.

Chapter 9

The BBC micro:bit and the Raspberry Pi

In this chapter

- An introduction to the Raspberry Pi educational microcomputer
- How to connect the BBC micro:bit to the Raspberry Pi and read from its internal sensors or add-on hardware
- How to use the BBC micro:bit as an external display device for the Raspberry Pi
- A practical example of Raspberry Pi-to-BBC micro:bit communication: a BBC micro:bit-powered Raspberry Pi CPU monitor

THE BBC MICRO:BIT is a powerful computing device in its own right, and there is no end of projects you can develop using only the BBC micro:bit and a few low-cost components. Pairing it up with the Raspberry Pi, however, opens a whole new world of potential for both the BBC micro:bit and the Raspberry Pi—and if you already have a Raspberry Pi and a BBC micro:bit, you don't need anything else to get started.

At its simplest, the Raspberry Pi can be used with the BBC micro:bit like any other computer. The BBC micro:bit appears as a removable drive, labelled MICROBIT, and you can write programs in the browser, download them, and drag them to the BBC micro:bit just as with a more expensive desktop or laptop. Delve a little deeper, however, and you can integrate the BBC micro:bit more fully into your Raspberry Pi programs, using it as an external display or reading from its various sensors—combining the two platforms into a powerful educational experimentation tool.

NOTE The Raspberry Pi runs a version of the GNU/Linux operating system dubbed Raspbian, which includes programming and other educational tools preloaded. The techniques described in this chapter can also be used on any other computer featuring the same tools, including your existing desktop or laptop. The easiest way to achieve this is to download and run Debian+Pixel, a GNU/Linux distribution for desktop and laptop computers from the Raspberry Pi Foundation. Running this, your existing computer will be able to do everything in this chapter just as though it were a Raspberry Pi.

More information on Debian+Pixel, which is available to download free of charge and is compatible with most computers from the last decade or so, can be found on `raspberrypi.org/blog/pixel-pc-mac`.

About the Raspberry Pi

Released in February 2012, the Raspberry Pi was an immediate success. Offering similar capabilities to a full-size desktop or laptop computer in a footprint no larger than a credit card, the Raspberry Pi has found a home in education and tinkerers' labs throughout the world, and it recently took its place in the history books as third best-selling computer ever.

The Raspberry Pi has a couple of major advantages over a traditional desktop computer, alongside its pocket-friendly size: it's cheap, costing no more than a single videogame or a couple of new-release films in its most expensive incarnation, and it features a 40-pin general-purpose input-output (GPIO) header for communicating with external hardware (see Figure 9-1)—much like the BBC micro:bit's 25-pin edge connector.

FIGURE 9-1: A Raspberry Pi 3 microcomputer

There are several models of Raspberry Pi, all of which are compatible with the BBC micro:bit. The smallest and cheapest, the Raspberry Pi Zero and wireless-equipped Raspberry Pi Zero W, require a *USB On-The-Go* (OTG) adaptor to connect to the BBC micro:bit's USB port; the larger models, including the compact Raspberry Pi Model A+, Raspberry Pi 2, and powerful Raspberry Pi 3, can connect with no additional hardware beyond the micro-USB cable you use with your desktop or laptop.

To continue with this chapter, set up your Raspberry Pi with the latest Raspbian operating system and power it on. If you're using a Raspberry Pi Zero, Raspberry Pi Zero W, or Raspberry Pi Model A+, you need a USB hub so you have space for your keyboard, mouse, and the BBC micro:bit's USB cable; if you're using the Raspberry Pi Model B+, Raspberry Pi 2, or Raspberry Pi 3, make sure you have one of its four USB ports free for the BBC micro:bit.

Connecting the Raspberry Pi to the BBC micro:bit

The Raspberry Pi is a fully-functional microcomputer that is capable—given enough time and resources—of doing anything its larger equivalents can do. Connecting the BBC micro:bit to the Raspberry Pi, then, is no different from connecting it to a Windows, macOS, or Linux PC, as described in Chapter 3, 'Programming the BBC micro:bit'.

With the Raspberry Pi switched on and the Pixel desktop visible, connect a micro-USB cable to one of the Raspberry Pi's free USB ports. If you're using a model with only one port, you need to use a USB hub so you have room for the keyboard and mouse as well. Connect the other end to the BBC micro:bit's micro-USB port, and after a few seconds, a dialog box offering to open the BBC micro:bit's removable drive in the File Manager application will load (see Figure 9-2). You can click the OK button to open the drive and drag an existing hex file to it, or you can click Cancel to close the dialog without launching the file manager.

If you want to simply program the BBC micro:bit using the Raspberry Pi, you can load the Chromium browser from the Raspberry Menu button at the upper left of the screen, visit `microbit.org`, and begin programming as described in Chapter 3. Everything works exactly as it does on your desktop or laptop, and the programs you download need only be dragged and dropped to the MICROBIT drive to flash and loaded.

An alternative to running the programming tools in your browser is to use *Mu*, which is designed specifically for programming the BBC micro:bit using the MicroPython language, and comes with tools to make that process easier. Mu, however, may not be installed by default, depending on the version of Raspbian you are using. To install it, make sure your Pi is connected to the Internet, open the terminal from the top menu bar or from the Accessories submenu of the Raspberry Menu icon at the top left of the screen, and type the following command:

```
sudo apt-get update && sudo apt-get -y install mu
```

FIGURE 9-2: Pixel's removable drive prompt

This refreshes the list of available packages and installs the latest version of Mu. Once Mu is installed, you can launch Mu from the Programming submenu of the Raspberry menu at the top left of the screen (see Figure 9-3).

FIGURE 9-3: The Mu integrated development environment

Designed with the needs of beginners in mind, Mu makes it as easy as possible to write programs for the BBC micro:bit. Any code which has been written in or for the browser-based MicroPython editor works unchanged in Mu, but the menu across the top offers extra functionality not available in the browser editor: Flash and Repl.

Note, however, that some functions—including support for using the BBC micro:bit's radio, as described in Chapter 8—may not be available in the version of Mu available in Raspbian. If your program relies on any of these functions, indicated by a syntax error displayed in Mu when the program works fine through the browser-based Python Editor, then you will either need to use the Python Editor (at `python.microbit.org`) or try the latest version of Mu from `codewith.mu`.

TIP

When using Mu, clicking the Flash button after saving your program takes care of loading it onto the BBC micro:bit. Instead of having to open your file manager, find the saved program, and then drag it to the MICROBIT drive, you can simply click Flash and the program is automatically installed onto the BBC micro:bit. If you change your program and want to update the BBC micro:bit, simply click the Save button and the Flash button again; you can repeat this as many times as you like.

Clicking the Repl button opens an *interactive shell*, known as a *Read-Eval-Print-Loop* (REPL), which allows you to type MicroPython code to be run directly on the BBC micro:bit (see Figure 9-4). Instead of writing a list of instructions which the BBC micro:bit then runs one at a time, you can type a single instruction and have it immediately run on the BBC micro:bit— a time-saving way of testing out new instructions and features without having to go through the process of writing a full program, saving it, and flashing it to the BBC micro:bit.

FIGURE 9-4: The Repl interactive shell in Mu

When using the Repl shell in Mu, any instruction you type is run immediately after you press the Enter key. However, the usual rules of a Python or MicroPython program apply (as discussed in Chapter 7, 'Python'): you need to import instructions before you can use them, and if you're using loops or other nested code you need to pay attention to the indentations at the start of each line.

A simple way to test out Repl's capabilities is with the two-line 'Hello, World!' program from Chapter 7. Delete any program code from the main Mu window and click the Flash button to load MicroPython onto your BBC micro:bit, and then click the Repl button to load the shell and type the following two lines one after the other:

```
from microbit import *
display.scroll('Hello, world!')
```

As soon as you press the Enter key to end the second line, you see the message `Hello, world!` scroll across the BBC micro:bit's LED display. When the message ends, so does the program. Without a loop, there's nothing to tell the BBC micro:bit what to do next. You can type any instructions in this way—even creating an infinite loop to keep the program running—but for a program of any more than a few lines it's much easier to write them in an editor, either Mu or the browser-based Python Editor, and flash them to the BBC micro:bit.

Mu is a useful tool for micro:bit programming in MicroPython, but you can also communicate with the BBC micro:bit in another way: integrating it into programs running on the Raspberry Pi, or integrating the Raspberry Pi into programs running on the BBC micro:bit. The remainder of this chapter looks at some of the possibilities that open up when you use the two devices in harmony.

Reading from the BBC micro:bit

The BBC micro:bit has a number of features an unexpanded Raspberry Pi lacks, including the 5×5 LED display, accelerometer, magnetic compass, and analogue-to-digital (ADC) converters attached to its pins (see Chapter 10, 'Building Circuits', for more on these). All of these features can be used from the Raspberry Pi over the micro-USB cable, using the full version of the Python programming language.

To get started, you first need to find out the *device name* the BBC micro:bit is using—the name given to the device under the */dev* directory in Linux. Usually, this is */dev/ttyACM0*; in some cases, especially if you have other devices connected to your Raspberry Pi, the number at the end of the device name differs. To find your device name, connect the BBC micro:bit to the Raspberry Pi. If the MICROBIT drive pops up, close that window and open the terminal from the Raspberry menu at the upper left of the screen before typing the following command:

```
dmesg | tail -20 | grep ttyACM
```

If your BBC micro:bit is properly connected to your Raspberry Pi, the above command returns its device name (see Figure 9-5); if it doesn't, check that it is properly connected and that you're using a working micro-USB cable with both power and data pins connected. If the device name ends in a number other than 0, make sure to change the following code to match your BBC micro:bit's device name; otherwise, you can use them as they are.

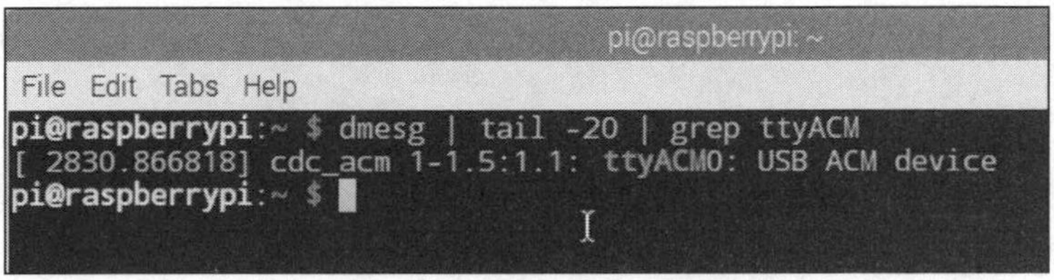

```
pi@raspberrypi: ~
File  Edit  Tabs  Help
pi@raspberrypi:~ $ dmesg | tail -20 | grep ttyACM
[ 2830.866818] cdc_acm 1-1.5:1.1: ttyACM0: USB ACM device
pi@raspberrypi:~ $
                        I
```

FIGURE 9-5: Discovering the BBC micro:bit's device name

This is called a *serial device*, and it is different from the BBC micro:bit's removable drive. Where the removable drive allows you to drag a completed program onto the BBC micro:bit for flashing (see Chapter 3), this device allows you to send and receive data from whatever program is currently running on the BBC micro:bit.

To read data from the BBC micro:bit, you need to have it running its own program and sending the results to the serial device. First you need to set up the BBC micro:bit so it's constantly taking readings from one of its sensors or pins. We'll use the accelerometer as an example, but you can easily modify the following program to read from the compass, the buttons, or any hardware connected to the BBC micro:bit's pins. Open the MicroPython editor in your browser or the Mu editor and type in the following (see Figure 9-6); in either editor, the first line is already typed in for you and is here for completeness.

> **TIP**
> When a line of code would extend past the border of the page, a ↵ symbol is printed. When you see this symbol, continue to type the code without pressing the Enter or Return keys. If you're not sure how a line of code should be entered, visit the website at `www.wiley.com/go/bbcmicrobituserguide` to download plain-text versions of each program; these can then be used for reference or even simply copy and pasted directly into the editors.

```
from microbit import *
while True:
    x, y, z = accelerometer.get_x(),↵
  accelerometer.get_y(), accelerometer.get_z()
    print(x, y, z)
    sleep(500)
```

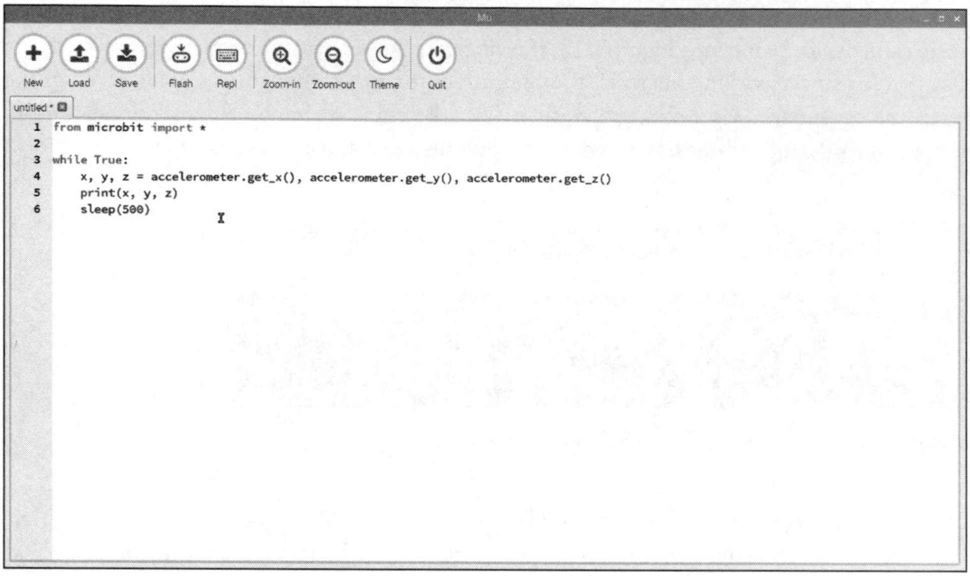

FIGURE 9-6: The accelerometer program in Mu

The `accelerometer.get` instruction is explained in Chapter 7, but the `print` instruction is new. Ordinarily, this instruction tells Python to print something—in this case the x, y, and z variables which hold the position data from the accelerometer—to the console, and if you're running Repl, that's exactly what they do. Running without Repl, though, there's no console, so the `print` instruction instead outputs via the BBC micro:bit's serial port. That's something we can read on the Raspberry Pi.

Flash the program onto the BBC micro:bit, either by saving it from the browser-based editor and dragging it to the MICROBIT drive or by clicking the Save button in Mu, naming the program 'accelerometerreader.py', and then clicking the Flash button. When the process has finished, it's ready to test. Open the terminal from the Raspberry menu at the top left of the screen and type the following:

```
sudo apt update && sudo apt install -y screen
```

This command will update the list of software available for your Raspberry Pi and then install a utility called `screen`. This tool, among its other features, allows you to connect to serial devices like the BBC micro:bit to send and receive data. To connect `screen` to your BBC micro:bit, close Mu and type the following in the terminal:

```
screen /dev/ttyACM0 115200
```

Remember to change the number at the end of the device name if necessary. This `screen` command opens a *serial console* to the BBC micro:bit, and you should start to see the terminal fill with data from the accelerometer (see Figure 9-7). Try picking up the BBC micro:bit and moving it around to change the numbers, being careful not to disconnect the micro-USB cable at either end.

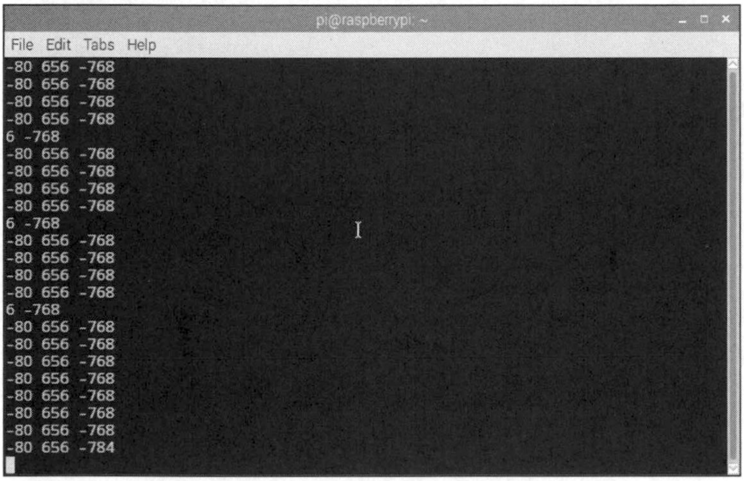

FIGURE 9-7: Reading accelerometer data via screen

You're not limited to simply displaying the data in the terminal, either. Using any programming language on the Raspberry Pi that comes with the ability to communicate with serial devices, you can receive and send data from and to the BBC micro:bit—allowing you to integrate it into your own programs.

Open the Python 2 (IDLE) editor from the Programming submenu of the Raspberry menu, click on File and then New File, and type in the following program:

```
import serial
ser = serial.Serial("/dev/ttyACM0", 115200, timeout=1)
ser.close()
ser.open()
while True:
    accelerometerData = ser.readline()
    print(accelerometerData)
```

This is about the simplest example of a Python program to read from a serial port as possible. After importing the `serial` library on the first line and configuring it on the second— remembering to change the BBC micro:bit's device name if necessary—the program closes the port in case any other program was using it and then opens it for its own use. The program then enters a loop, reads a single line from the BBC micro:bit's serial port, prints it to the screen, and then loops back to wait for the next line.

Click on File and then Save and save the program as 'accelerometerprint.py'; then click Run Module from the Run menu to launch it. The Python console begins to fill with readings from the accelerometer. Try moving the BBC micro:bit around to witness the values change (see Figure 9-8).

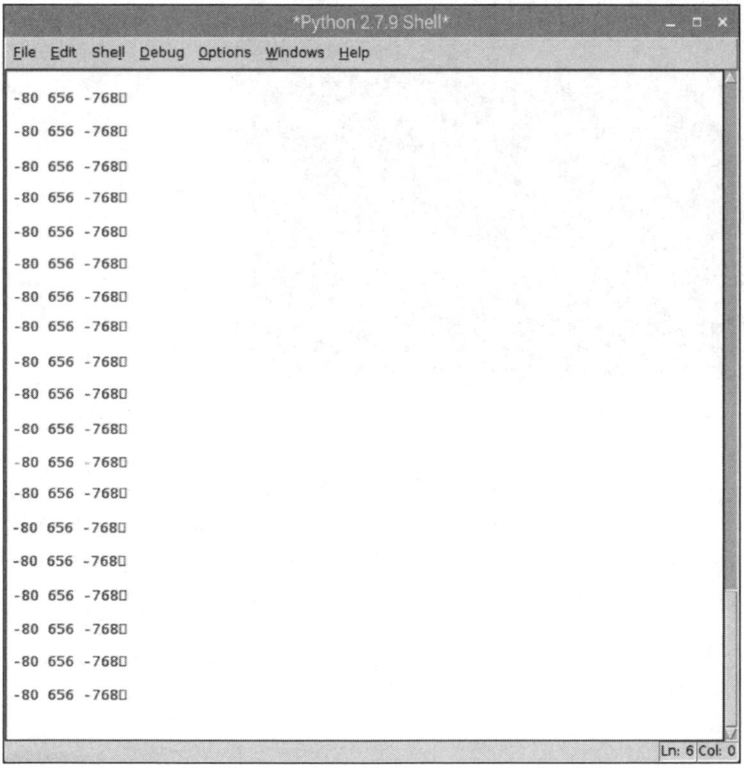

FIGURE 9-8: Reading accelerometer data via Python

Line by line, the program does the following:

```
import serial
```

This tells Python to load the `serial` library. You may notice that you're not importing the usual `microbit` library here; that's because this code is running on the Raspberry Pi rather than the BBC micro:bit, so the only library we need is to handle the communication between the BBC micro:bit and the Raspberry Pi.

```
ser = serial.Serial("/dev/ttyACM0", 115200, timeout=1)
```

This line creates an object called `ser` using the `serial` library, with three parameters. The first parameter, `"/dev/ttyACM0"`, is the device name for the BBC micro:bit's serial connection and should be changed if you found a different name earlier in the chapter. The second parameter is the *bitrate* or *baudrate* of the serial connection, which for the BBC micro:bit should be set to 115,200 bits per second (bps). The final parameter sets a one-second timeout on reads from the serial device, so the program doesn't get stuck.

```
ser.close()
ser.open()
```

These two lines work together. The first closes the serial device in case it had been left open during an earlier run of the program while you were testing things. The second opens it, ready for communication between the BBC micro:bit and the Raspberry Pi.

```
while True:
```

As always, this line begins the infinite loop that stops the Python program from exiting as soon as it reaches the end of the instructions. The lines under this need to be indented by four spaces so that Python knows they belong inside the loop.

```
    accelerometerData = ser.readline()
```

This line uses the `serial` library to read a single line from the BBC micro:bit, which, when it is running the program you wrote earlier in the chapter, contains data from the accelerometer. This line is stored in the variable `accelerometerData` for use later in the program.

```
    print(accelerometerData)
```

Finally, this line prints the `accelerometerData` variable to the console, making it appear on-screen.

For more information on using data from the BBC micro:bit as variables in your own Python programs, please consult a general Python text, such as the user-friendly 'Python for Beginners', available for free at **www.python.org/about/gettingstarted/**.

Using the BBC micro:bit Display

The BBC micro:bit's serial port is *bidirectional,* meaning that it's possible not only to read from it but also to write to it. Using this, you can send data to the BBC micro:bit as well as receive data from it, allowing your Raspberry Pi to take a more active role in controlling the way the program runs.

While the above example requires a dedicated program to be running on the BBC micro:bit, constantly checking its sensor readings and printing them to the serial device, sending commands to the BBC micro:bit from the Raspberry Pi is best achieved through the Repl console.

To start, you need to flash the BBC micro:bit with the Repl console. This is a somewhat hidden feature of both the browser MicroPython editor and Mu, but it can be accessed in a simple way in both: just load up either editor but don't type in any program code (see Figure 9-9). Click Download and drag the resulting hex file to the MICROBIT drive if using the web editor, or click Save and then Flash if you're using Mu.

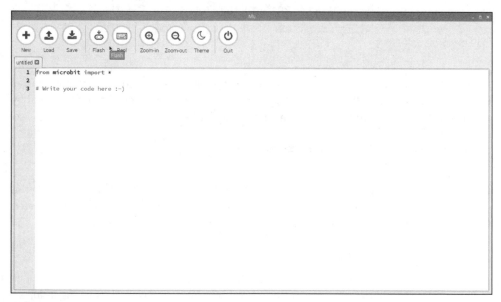

FIGURE 9-9: A blank MicroPython program in Mu

This special hex file contains the MicroPython Repl console, which is accessible using the serial device. To actually use this console to read accelerometer data, click on Python 2 (IDLE) from the Raspberry menu at the top left of your Raspberry Pi desktop, click File and then New, and then type in the following program (see Figure 9-10).

Remember to enter all lines with a ↵ as one line.

```
import serial, time
ser = serial.Serial("/dev/ttyACM0", 115200, timeout=1)
ser.close()
ser.open()
ser.write("from microbit import * \r".encode())
while True:
    ser.write("display.scroll('Hello, world!')↵
  \r".encode())
    time.sleep(10)
```

FIGURE 9-10: A display-writing program in IDLE

Click on File and then Save, and save the program as 'displaywrite.py'. Then click Run Module from the Run menu to launch it. This program is a little different from the ones you may have written for the BBC micro:bit itself, so we'll look at it line by line.

```
import serial, time
```

This line tells Python to import the `serial` and `time` libraries, so you can send instructions to the BBC micro:bit's serial device and slow down the loop process.

```
ser = serial.Serial("/dev/ttyACM0", 115200, timeout=1)
```

As before, this creates a `ser` object using the `serial` library set to the BBC micro:bit's serial device name and bitrate.

```
ser.close()
ser.open()
```

These two instructions ensure that the port was cleanly closed if it has been used previously, then open the serial device ready for your instruction to be sent.

```
ser.write("from microbit import * \r".encode())
```

When communicating with Repl, you always need to begin by importing the `microbit` library just the same as if you were using it directly or writing a program in one of the MicroPython editors. Using it via the serial device is no different, and this line tells the Python `serial` library to send the instruction `from microbit import *` to Repl. The final `\r` at the end of the instruction sent to the BBC micro:bit acts like the Enter key on the keyboard, while the `.encode()` instruction formats the message in a way that the `serial` library understands.

```
while True:
```

This begins the program's infinite loop. By putting the previous instruction to import the `microbit` library above this line, we ensure it only runs once. There's no need to re-import the library every time the loop restarts.

```
    ser.write("display.scroll('Hello, world!') ↵
  \r".encode())
```

Remember to enter all lines with a ↵ as one line. This line uses the `ser.write` instruction again, this time sending the command `display.scroll('Hello, world!')` to the BBC micro:bit to make what should by now be a familiar message scroll across the display. As before, the instruction sent includes the `\r`—or *carriage return*—to simulate someone pressing the Enter key and telling MicroPython to execute the command.

```
time.sleep(10)
```

Finally, this line waits ten seconds before returning to the start of the loop and displaying the message all over again.

Using this basic template, you can send any instruction or list of instructions to the BBC micro:bit. For instance, replace the contents of the loop with the following to have the BBC micro:bit show a smiley face for ten seconds, a sad face for ten seconds, and then go back to the smiley face again:

```
ser.write("display.show(Image.HAPPY) \r".encode())
time.sleep(10)
ser.write("display.show(Image.SAD) \r".encode())
time.sleep(10)
```

Replace the contents of the loop with the following to toggle Pin 0 on for ten seconds and then off for ten seconds in a repeating pattern. Connect an LED to Pin 0 to see the effect, as described in Chapter 10.

```
ser.write("pin0.write_digital(1) \r".encode())
time.sleep(10)
ser.write("pin0.write_digital(0) \r".encode())
time.sleep(10)
```

Practical Example: A CPU Monitor

A good example of how the BBC micro:bit can be put to use with your Raspberry Pi is to turn its display into a graph of the Raspberry Pi's CPU load—a measure of how hard the processor is working. Sat at the desktop, you see short bars on the graph; play a game or load a big web page, and the bars shoot up as the processor gets to work.

This sample project uses the MicroPython Repl console, so if you're running a different program on your BBC micro:bit, you need to flash a copy by clicking Download on an empty MicroPython program in the web editor and dragging it to the MICROBIT drive or clicking the Save and Flash buttons in Mu.

Before writing the CPU monitoring program itself, you need to install a new Python library called `psutil` (process utilities). To do so, load the terminal by clicking on the Terminal icon on the top menu bar or its entry in the Accessories submenu of the Raspberry menu, and type the following command:

```
sudo apt-get -y install python-psutil
```

To write the Python program, click on the Raspberry menu, choose Programming, and click Python 2 (IDLE) from the menu. When IDLE loads, click the File menu, and then New File, and type in the following program (see Figure 9-11), remembering to continue typing on the same line whenever you see the ↩ symbol.

This program is quite dense and can be difficult to enter by hand, particularly when it comes to the line which draws the bar graph itself. Remember to keep typing when you see the ↩ symbol, only pressing Enter or Return when you get to the end of a line without the symbol. If you find you're getting syntax errors, double-check your program against the book, paying close attention to things like the number of brackets and other symbols; alternatively, you can download a copy of the program from www.wiley.com/go/bbcmicrobituserguide.

```
import serial, psutil, time
gradients = 20
readingList = [0,1,2,3,4]
ser = serial.Serial("/dev/ttyACM0", 115200, timeout=1)
ser.close()
ser.open()

print "Started monitoring system statistics for↩
 micro:bit display."
ser.write("from microbit import * \r".encode())
time.sleep(0.1)
ser.write("display.clear() \r".encode())
time.sleep(0.1)

barGraph = [[0, 0, 0, 0, 0], [0, 0, 0, 0, 0], [0,↩
 0, 0, 0, 0], [0, 0, 0, 0, 0], [0, 0, 0, 0, 0]]

while True:
    sysLoad = psutil.cpu_percent(interval=0)
    readingList.insert(0,int(sysLoad))
    del readingList[5:]
    for x in range(5):
        for y in range(5):
            readingComparison = (y+1) * gradients
            if (readingList[x] >= readingComparison):
                barGraph[y][x] = 9
            else:
                barGraph[y][x] = 0
    ser.write("BARGRAPH = Image↩
(\"%s:%s:%s:%s:%s\") \r".encode() % ↩
(''.join(str(e) |for e in barGraph[0]), '↩
```

```
'.join(str(e) for e in barGraph[1]), '↵
'.join(str(e) for e in barGraph[2]), '↵
'.join(str(e) for e in barGraph[3]), '↵
'.join(str(e) for e in barGraph[4])))
    time.sleep(0.1)
    ser.write("display.show(BARGRAPH) \r".encode())
    time.sleep(0.9)
```

FIGURE 9-11: The CPU monitoring program in IDLE

Click the File menu, click Save, and save the program as 'cpumonitor.py' before clicking Run Module from the Run menu to execute it. After a second or so, you see LEDs begin to light up on the BBC micro:bit display as the program graphs your CPU load. Try loading a complex website, a game, or playing a video to see the load increase; when you go back to idling at the desktop, the load decreases accordingly (see Figure 9-12). Each LED in the vertical bars represents 20% CPU load, and if all five LEDs in a bar are lit, your Raspberry Pi is using 100% of its CPU. Each column of LEDs represents one second, with the bars scrolling sideways as they update.

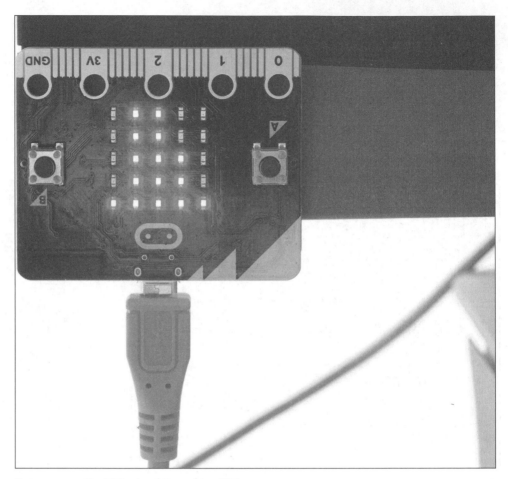

FIGURE 9-12: The BBC micro:bit graphing CPU usage

You can use the same program to graph various Raspberry Pi system resources. Try replacing the following lines:

```
sysLoad = psutil.cpu_percent(interval=0)
readingList.insert(0,int(sysLoad))
```

with these lines:

```
sysMem = psutil.virtual_memory()
readingList.insert(0,int(sysMem.percent))
```

For more information on the statistics you can graph using the psutil Python library, visit pythonhosted.org/psutil/.

Chapter 10
Building Circuits

In this chapter

- An introduction to the basic components of electronic circuits
- The BBC micro:bit's input-output pins and what they can do
- Using the BBC micro:bit to power your electronic circuits
- How to read from and write to the BBC micro:bit's pins from your own programs
- Reading from a button input
- Writing to an LED output
- A practical project: building your own set of traffic lights

THE BBC MICRO:BIT'S on-board buttons and LED display make it a device which works perfectly well on its own. You can build surprisingly complex projects without ever adding additional hardware, from games to touch detectors.

Sooner or later, though, you're going to want to expand the capabilities of your BBC micro:bit, and it's here the *edge connector* comes into play. This connector includes multiple *input-output pins* which can be used to talk to your own circuits.

Before you get started, though, you need to learn a little bit about the equipment, components, and terminology that go into making electronic circuits.

Electronic Equipment

Before you can begin building electronic circuits to connect to the BBC micro:bit, you need some basic equipment. If you purchased the BBC micro:bit on its own, you need to find these separately; if you purchased an expansion pack or experimenter's kit, you should find the equipment you need included in your bundle.

The following is a basic list of what you need to get started with building the sample projects later in this chapter. As you begin experimenting with your own circuit designs, you need additional components, but the following is enough to get you started.

- **Wires with crocodile clips or banana plugs**—The BBC micro:bit is designed to be easily connected to additional hardware using *jumper wires* which end in *crocodile clips* (sprung metal clips with serrated teeth which look a little like the jaws of a crocodile) or *banana plugs* (split poles which widen in the middle to grip the edges of a connector). You can use either type of connector, but make sure you buy the right size. Crocodile clips should be no wider than 4mm, though thinner will work; banana plugs must be 4mm in diameter exactly, or they're too loose or too tight to fit through the holes. Try to buy bundles that come in multiple colours, too, as it makes your layout considerably easier to follow as you're building the circuit.

- **Buttons or switches**—The BBC micro:bit has two buttons on board, but an external button or two is a great way to learn about reading basic input signals. Almost any button or switch works, but ideally look for *momentary switches* which send a signal only while being actively held down; these are the same switches as you find on the BBC micro:bit itself.

- **Light emitting diodes (LEDs)**—An LED is a fantastic and inexpensive way to quickly see whether your program is sending a signal or not. LEDs come in all shapes and sizes, but for your first few projects keep things simple and buy single-colour LEDs suitable for low-current use. Avoid extremely bright white or blue LEDs; these typically require more current than the BBC micro:bit's pins can provide and need an external power supply and an additional component, called a *transistor*, to switch them on and off.

- **Resistors**—Most electrical circuits use *resistors* as a means of controlling the voltage or current flowing through a component. Resistors are particularly important when working with LEDs. If you wire an LED directly into a BBC micro:bit pin without a resistor to limit the current, it's theoretically possible to burn the LED out. Try to buy a selection of different values of resistor, measured in *ohms* (Ω). A pack with 10KΩ and 68Ω resistors is enough to see you through most BBC micro:bit projects.

- **Potentiometers**—Also known as *varistors*, or *variable resistors*, *potentiometers* are resistors, the value of which can be changed by turning a knob or screw. The most common potentiometers have a maximum 10KΩ resistance and are often used in projects as volume or brightness controls.

The BBC micro:bit is designed to be used without another form of equipment, and circuits can be built using nothing more than the wires and components. If you're looking to make your project more permanent, however, you may need to invest in a few additional pieces.

- **Breadboard**—Also known as a *solderless breadboard*, the breadboard is a plastic tray covered in small holes. Wires and the legs of components are inserted into these holes, and hidden metal strips on the underside of the breadboard join them in an electrical circuit. A breadboarded circuit has the benefit of being less liable to come apart than one made with wires and crocodile clips, while still allowing you to remove the components and reuse them for a different project later. Look for a breadboard which features 2.54mm pin spacing.

- **Breadboard jumper wires**—Crocodile clips are of little use with a breadboard, so you need a selection of breadboard jumper wires which end in either plain wire or soldered pins. Remember that you still need to connect the breadboard to the BBC micro:bit, so make sure you have a few wires with a crocodile clip or banana plug at one end to make those connections; alternatively, buy a strip of *2.54mm male pin headers* and insert them into the breadboard to give you something to attach a crocodile clip to.

- **Stripboard**—The next step up from a breadboard, a *stripboard* allows you to make permanent circuits by soldering components into place. Like a breadboard, the stripboard has rows of metal underneath which make up the circuit; these rows can be cut using a *track cutting tool* (or a small drill bit) to alter the shape of the circuit.

- **Soldering iron**—If you're using a stripboard, you need a *soldering iron,* a screwdriver-like device which heats up a metal tip to melt metal solder and create permanent electrical connections between the components and the stripboard. Make sure you choose a soldering iron which is labelled as suitable for electronics work and with a reasonably fine tip, and always remember that soldering irons get extremely hot. Try to find an iron which comes with a basic accessories kit including a stand, some electrical solder, and a sponge or wire ball for cleaning the tip of the iron.

- **Side cutters**—The components you use to make your circuit are known as *through-hole technology* (THT) components, characterised by their long legs. If you're building a stripboard circuit, you're going to need to cut those down to size when you've finished soldering them into place; to do this, you need a pair of *small side cutters*.

- **Multimeter**—A *multimeter* is a multifunction measuring device, which at its most basic is capable of measuring voltage, amperage (current), and resistance. It's particularly useful if you need to diagnose a problem with your circuit. While professional versions can be extremely expensive, even the cheapest models are suitable for hobby projects.

As your skills grow and you look to build bigger and more complex projects, you may find you need additional tools, such as tweezers, a work stand (also known as *helping hands*), or tools for removing rather than applying solder. You may also find a *break-out board,* which makes it easier to use the BBC micro:bit with a breadboard or stripboard, a useful accessory; more information on these can be found in Chapter 11, 'Extending the BBC Micro:bit'.

For now, you can begin learning with just the parts from the first list.

The Input-Output Pins

The edge connector on the BBC micro:bit takes the form of copper *pads* which are connected to the processor via electrical *traces* on the circuit board (see Figure 10-1). Although they're flat, it's traditional to refer to these pads as *pins* as they connect to the pins of the processor.

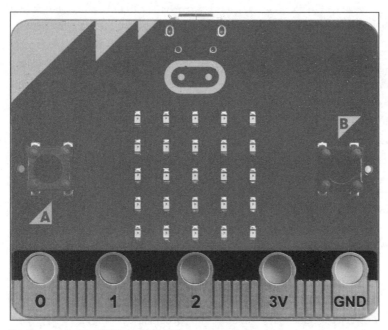

FIGURE 10-1: The BBC micro:bit edge connector

There are 25 pins in total, split into five large pins and 20 smaller pins. The large pins are labelled with the numbers 0, 1, and 2, for the three general-purpose input-output pins, 3V for the positive voltage pin which provides power to your circuit, and GND for the ground, or negative, pin (see Figure 10-2). The smaller pins aren't labelled, but they are referred to using the numbers 3 through 22 to avoid mixing them up with the three larger pins.

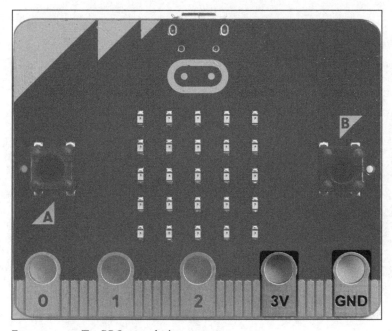

FIGURE 10-2: The BBC micro:bit's power pins

Always remember that the BBC micro:bit is a sensitive electronic device. Never connect or disconnect hardware from its pins while the power is on, and always check and double-check connections before connecting power. While it's designed to be robust, the BBC micro:bit can be damaged if wired incorrectly.

WARNING

The Large Pins

The large pins are clearly labelled on the top of the BBC micro:bit board and are split into a group of three general-purpose input-output pins to the left and the two power pins to the right (see Figure 10-3).

The three pins labelled 0, 1, and 2 can be used as inputs or outputs depending on what hardware you're connecting to the BBC micro:bit. An LED, for example, is an output device and can be made to light up when connected to one of the three numbered pins; a button or switch is an input device, and the BBC micro:bit can be programmed to respond when it is connected to another of the pins—just like the on-board Button A and Button B.

The large pins are always available for you to use in your programs and do not interfere with any of the on-board switches, LEDs, or other hardware. You can use them to add additional inputs or outputs, such as adding a small speaker, more LEDs, or extra buttons.

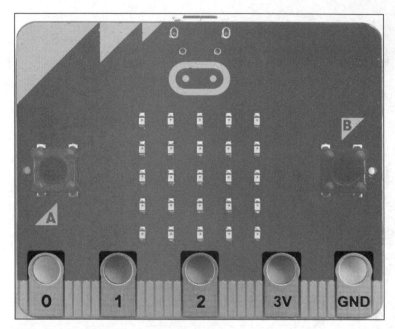

FIGURE 10-3: The BBC micro:bit's large pins

Each of the numbered large pins is connected to a shared *analogue to digital converter* (ADC), a part of the processor that allows it to read more than the usual on-off 0-1 binary of a switch. When it's connected to hardware that provides an analogue signal, such as a light sensor or a microphone, you can read smooth changes in the input—great for projects such as room alarms or sound-sensitive lighting. You can find out more about using the ADC feature later in this chapter.

The remaining two large pins are connected to the BBC micro:bit's power supply. The pin marked 3V is also known as the positive pin and can be used to power your circuits; the pin marked GND is the ground, or negative, pin, and it completes the circuit.

All five of the large pins are designed to be used with wires connected to crocodile clips or 4mm banana plug connectors. When using banana plugs, simply insert them into the hole on the relevant pin. If you're using crocodile clips, you need to be careful that you don't sit the clip over the smaller pins next to each large pin and that they don't shift and slide while you're using the BBC micro:bit. The easiest way to do this is to insert the clips so they are sticking vertically up out of the BBC micro:bit (see Figure 10-4), which stops them from touching other pins while also making them more secure against being knocked or accidentally disconnected.

FIGURE 10-4: Using crocodile clips with the BBC micro:bit

Whatever method you use to wire hardware to the BBC micro:bit, always check your connections before powering it on. Be especially careful with the 3V and GND pins; connecting these together without any hardware between them could result in a *short circuit* which could permanently damage the BBC micro:bit.

The Small Pins

As well as the five large pins, the BBC micro:bit has 20 additional smaller pins on its edge connector (see Figure 10-5). These pins aren't designed for use with crocodile clips or banana plugs, but instead provide the option to add hardware to the BBC micro:bit using a *female edge connector*. Add-on devices using these smaller pins are available from third-party companies and range from *break-out boards* which make it possible to use the additional pins with a breadboard to robots which use the BBC micro:bit as their 'brain'. More information on some of these add-ons can be found in Chapter 11.

Unlike the large pins, many of the BBC micro:bit's smaller pins are shared with on-board hardware. Pin 3, for example, is also used to control the first column of the BBC micro:bit's LED display; Pin 11, meanwhile, is connected to Button B. This doesn't stop you from using most of these pins in your own circuits, but it does mean that you can't use the corresponding on-board device at the same time—and in the case of the LEDs, you may find that some light or switch off when you're not expecting them to! To avoid this, you can disable the display using the [led enable false] block in JavaScript Blocks, led.enable(false) instruction in JavaScript, or the display.off() instruction in Python.

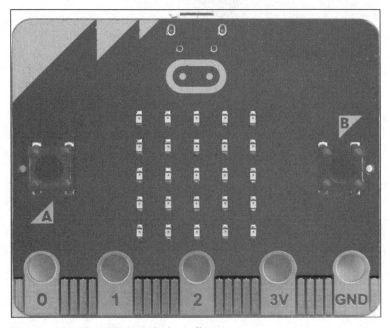

FIGURE 10-5: The BBC micro:bit's small pins

The small pins provide extra general-purpose input-output connections for more complicated projects, with three of the pins (Pin 3, Pin 4, and Pin 10) connected to the same ADC converter as the large pins. Some of these pins share functionality, with Pins 3 through 16, 19, and 20 being available for software control as general-purpose input-output pins. Additionally, some of the pins provide a means to communicate with more complex hardware through three special *buses*: Serial Peripheral Interface (SPI), Inter-Integrated Circuit (I^{2}C), and Universal Asynchronous Receiver/Transmitter (UART).

Finally, the pins at either side of the two power pins—3V and GND—are also power pins. By surrounding each large power pin with two smaller power pins, the chances of accidentally shorting the power pin to a general-purpose pin and damaging the BBC micro:bit are reduced—but if you're using crocodile clips, it's still possible to do so, so always check your connections.

TIP Full details on the large and small pins and the functions available on each can be found in Appendix D, "Pin-Out Listing".

The following two sections describe the SPI and I²C buses in general detail. These are advanced connections designed primarily for use with more complex add-on devices; if you're eager to learn about building your first project with simpler building blocks like switches and LEDs, skip the following two sections and continue with the rest of the chapter.

NOTE

Serial Peripheral Interface (SPI)

The Serial Peripheral Interface (SPI) bus is commonly used to talk to add-on hardware such as displays and keypads. It's available on Pins 13 through 15 of the BBC micro:bit edge connector. Pin 13 provides the *Serial Clock* (SCK) signal, Pin 14 the *Master Input Slave Output* (MISO) signal, and Pin 15 the *Master Output Slave Input* (MOSI) signal.

If you're connecting the BBC micro:bit to an SPI device, make sure you wire up the pins in the right order: SCK should go to SCK, MISO to MISO, and MOSI to MOSI on both devices. This is in contrast with other methods of serial communications where pins marked 'input' need to be connected to pins marked 'output' and vice versa. You can use multiple SPI devices with the BBC micro:bit by using GPIO pins as Chip Select (CS) pins, one pin per device.

For details on how to use the SPI bus, see the documentation that came with your SPI-compatible device.

Inter-Integrated Circuit (I²C)

The Inter-Integrated Circuit (I²C, pronounced 'I squared C') bus is typically used to talk to more complex add-on devices often powered by their own microcontrollers or microprocessors. These signals are available on Pins 19 and 20 of the BBC micro:bit edge connector. Pin 19 provides the *Clock* (SCL) and Pin 20 the *Data* (SDA) signals.

One of the major benefits of the I²C bus is that it allows for a large number of devices in a single circuit. Using I²C, it is possible to connect multiple slave devices to a single BBC micro:bit—or to connect multiple BBC micro:bits to a single I²C master device. The accelerometer and magnetometer both connect to the I²C bus and cannot be disabled. As a result, the I²C pins cannot be reused as general-purpose input-output (GPIO) pins even if you are not using the I²C bus in your own programs.

As with the SPI bus, if you're looking to connect an I²C device to the BBC micro:bit, you should consult the documentation that came with your hardware for instructions.

Universal Asynchronous Receiver/Transmitter (UART)

The BBC micro:bit's UART is responsible for handling the serial connection, but it can also be reassigned to use a pair of the edge connector pins: one pin to transmit data, and the other pair to receive data. This allows the BBC micro:bit to communicate with external devices that have their own UART, including other BBC micro:bits.

When reassigning the UART in a program, though, be aware that it will affect any other serial communications already configured. Assigning the UART to edge-connector pins will disable the BBC micro:bit's serial-over-USB capabilities, though this can be quickly restored by flashing a different program onto your BBC micro:bit.

Details on which pins can be used as UART receive and transmit pins can be found on the BBC micro:bit's edge connector data sheet at `tech.microbit.org/hardware/edgeconnector_ds/`.

TIP The power provided by the BBC micro:bit's pins depends on the power you provide to the BBC micro:bit itself. When running from USB, the BBC micro:bit will run at its peak voltage; when running from a battery pack, the voltage put out on each pin will drop as the charge in the battery pack drops. This can make parts like LEDs appear dimmer or motors run more slowly. Eventually, circuits may stop working until you replace the batteries or switch to a USB connection for power.

Your First Circuits

Now that you know where to find the BBC micro:bits pins and what they can do, it's time to start building your first circuits. The two circuits in this section of the chapter are as simple as they get and demonstrate the two main uses of the BBC micro:bit's pins: as inputs and as outputs.

The examples in this chapter are programmed using the Python language (see Chapter 7, 'Python'). The same circuits can be used with any of the other languages supported by the BBC micro:bit, however, and writing your own programs in the other languages is a great way to flex your skills.

WARNING When building these circuits, always use the BBC micro:bit to power the circuit as shown, and never try to make the circuit power the BBC micro:bit by connecting an external power supply or battery. While it's possible to build a more advanced circuit which supplies the 3V the BBC micro:bit needs, it's also easy to make a mistake and supply too much voltage which will damage the BBC micro:bit.

TIP The programs in this chapter are written in Python. If you would prefer to use JavaScript Blocks or JavaScript, versions written in these languages can be found in Appendix A and Appendix B, respectively.

Reading from a Button Input

The simplest possible circuit that actually achieves anything is a button, or switch, input. To build one, you need the following components from your collection:

- Wires with crocodile clips or banana plugs
- A button or switch

If you don't have a switch, you can simulate one by touching the ends of the two crocodile clips together for a 'press' of the switch and moving them apart for a 'release' of the switch.

Put your circuit together using the following instructions so it matches Figure 10-6, making sure that if you're using crocodile clips, they are attached to the BBC micro:bit securely and not accidentally bridging any of the smaller pins and beginning with the BBC micro:bit switched off.

1. If your switch has four legs, you're only going to need two of them. Position the switch so that it has two legs coming from its left side and two legs coming from its right side. For the following instructions, use one leg from the left and one from the right.

2. Attach a crocodile clip to the left leg of the switch; then attach the other end to Pin 0 on the BBC micro:bit using a crocodile clip or banana plug.

3. Attach a final crocodile clip wire to the right leg of the switch; then connect the other end to the 3V pin on the BBC micro:bit using a crocodile clip or banana plug to complete the circuit.

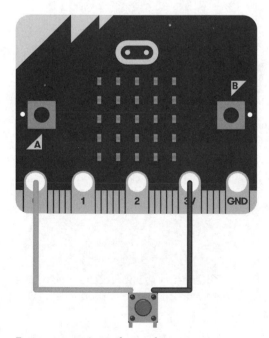

FIGURE 10-6: A simple switch circuit

<table>
<tr><td>TIP</td><td>A switch with four legs, rather than two, can trip you up. Your circuit will work only if you connect the two wires to a switched pair of legs; if you connect the two wires to legs on the same side of the switch, which are connected internally, the circuit will act as though you're constantly pressing the button.</td></tr>
</table>

If you can't tell how the legs are paired, you can test with a continuity tester or multimeter in continuity mode: place one probe on one leg then touch another leg with it; if the tester shows the two legs are connected, switch one probe to a different leg until you find two that aren't connected.

Alternatively, just go ahead and wire up the circuit; if the BBC micro:bit acts as though the button is being pressed even when it isn't, move one of the two wires to a different leg to fix it.

The BBC micro:bit now needs a program that tells it to read the switch as an input on Pin 0, which you need to write and upload to the BBC micro:bit. Load the Python editor in your web browser and enter the following program code:

```
from microbit import *
while True:
    while (pin0.read_digital() == 1):
        display.show(Image.SURPRISED)
    display.show(Image.ASLEEP)
```

Remember that indented lines in Python should always have four spaces for each indent (see Chapter 7 for more details); in this program, everything after the line `while True:` should be indented once with four spaces, except for the fourth line, which is indented twice with eight spaces total.

After double-checking all of your wiring, connect the BBC micro:bit to your computer with a micro-USB cable, click on the Download button, and drag the hex file to your MICROBIT drive to flash the program. The BBC micro:bit's display should change to a sleeping face. Press and hold the button to wake it up; then let go to send it back to sleep again.

Step by step, this program does the following:

```
from microbit import *
```

This line, as with all Python programs written for the BBC micro:bit, imports the code necessary to address the BBC micro:bit. Without it, the program doesn't work. For more details on `import` and the `microbit` module, see Chapter 7.

```
while True:
```

This line places the BBC micro:bit into an infinite loop, making it run the following instructions forever. Without it, the BBC micro:bit runs through the list of instructions just once before stopping. The only way to make it run through them again is to reset the BBC micro:bit; with this loop, the BBC micro:bit simply runs the program forever—or at least until you remove its power supply.

```
while (pin0.read_digital() == 1):
```

This second loop tells the BBC micro:bit what to do when the button is pressed. The `pin0.read_digital()` instruction tells the BBC micro:bit to read its Pin 0. If the button is being pressed, the read returns a value of 1; if it isn't, the read returns a value of 0. The `== 1` part of the instruction tells the BBC micro:bit to enter the loop only if the button is pressed; otherwise, the loop is skipped.

```
display.show(Image.SURPRISED)
```

If the BBC micro:bit has entered the loop—in other words, if you're pressing the button—this line runs next, showing a surprised face on the BBC micro:bit's display. The loop runs for as long as you hold down the button, keeping the surprised face on the display.

```
display.show(Image.ASLEEP)
```

When you release the button, or if you haven't pressed the button at all, this line tells the BBC micro:bit to show a sleeping face. By switching between these two images, it's easy to see if your program is reading your button presses correctly.

Reading Resistor Colour Codes

Before you can build the next circuit, you need to know how to tell the different values of resistor apart. Thankfully, most through-hole electronic components are clearly labelled: *capacitors* have their *capacitance*, measured in *farads*, printed on their casing, while *crystals* have their *frequency* likewise marked. Resistors, too, are marked with their values, but not in numbers; instead, they use a colour code.

The surface of a through-hole resistor is covered in coloured bands, with each band giving you one of the numbers you need in order to know both the value of the resistor in ohms and its *tolerance*—a value which tells you how close to the 'perfect' resistance value the resistor is likely to be. To decode these, use the diagram in Figure 10-7; the same diagram is available in full colour on the *BBC Micro:bit User Guide* website at `www.wiley.com/go/bbcmicrobituserguide`.

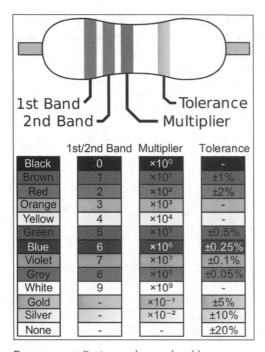

FIGURE 10-7: Resistor colour code table

Start by reading the value of the printed example resistor. The first two bands starting from the left indicate the resistance, while the next band indicates the *multiplier* and the final band indicates the tolerance. For the sample resistor, the first two bands are coloured red. On the table included in Figure 10-7, red equates to the value 2, so the initial number for the resistance value is 22.

The next band is the multiplier band and in the example is coloured green, which translates to 10^5 or 100,000 (1 followed by five zeros). As the name suggests, the multiplier is used with the initial resistance value: multiplying 22 by 100,000 equals 2,200,000, which is the true resistance value of the example resistor in ohms: 2,200,000Ω.

That's a big number, so it pays to simplify. There are 1,000 ohms in a kiloohm and 1,000 kiloohms in a megaohm. 2,200,000Ω is typically written as 2.2MΩ, or 2.2 megaohms.

The final band, located on the right side of the resistor, indicates the tolerance, or accuracy—in other words, it's how close to the listed resistance figure you would see were you to measure the resistor. For our sample resistor, this band is coloured gold, which indicates plus or minus 5 per cent. In other words, the actual resistance of the 2.2MΩ resistor could be as low as 2.09MΩ or as high as 2.31MΩ. The tighter the tolerance, the smaller that variation gets. For hobbyist use, though, almost any tolerance value gets you sufficiently close for the circuit to work.

Some resistors have five, rather than four, coloured bands around their bodies. To read these, simply treat the first three bands as the resistance values, the fourth as the multiplier, and the fifth as the tolerance. If you're unsure which way around the resistor goes to read it from left to right, remember that the tolerance band uses a different colour scheme and should always be to the right side.

Writing to an LED Output

The next circuit to build flips things around. Rather than using a pin as an input and reading from a button, it uses a pin as an output and lights up an LED. Remember that all three pins can act as either an input or an output, as what they do is defined entirely by the program you write.

To build this circuit, you need the following:

- Wires with crocodile clips or banana plugs
- An LED
- A suitable resistor

If you purchased a component kit with your BBC micro:bit, you can find resistors that are matched to the needs of the LEDs. If you're buying your components individually or have a kit of LEDs, you need to match your resistor to your LED to protect both the LED and the BBC micro:bit. In this circuit, the resistor acts as a *current limiting resistor*.

If you have a resistor already matched to your LED, you can skip the following explanation; otherwise, it tells you how to calculate exactly what resistor you need to run your LED at peak brightness without running the risk of damaging it.

To find the value of a current-limiting resistor that you need to protect an LED, you must know two things about the LED: its *forward voltage*, which is the voltage at which the LED is supposed to light, and its *forward current*, which is the maximum current the LED can draw before being damaged. This information can be found in the data sheet available from your component supplier.   **NOTE**

For the BBC micro:bit, always use LEDs with forward voltages given at 3.3V or 3V; any higher, and you need an external power supply and a transistor to switch them on and off. It's possible to use LEDs with forward voltages below 3V by using a higher-value resistor to protect them.

To calculate precisely what value of resistor you need for a given LED, you need to do some mathematics. The formula $R = (V - F) / I$ tells you what you need to know: R is resistance in ohms, V is the voltage applied to the LED, F is the forward voltage of the LED, and I is the maximum forward current of the LED in amps. Most LEDs have their forward current measured in milliamps (mA), so you need to divide the value by 1,000: a 25mA forward current is the same as 0.025A.

A typical red LED has a forward current of 25mA and a forward voltage of 1.7V. Knowing that the BBC micro:bit uses 3V power, you can now calculate the precise value of resistor required to protect the LED: `(3 - 1.7) / 0.025 = 52`, or 52Ω. Other LEDs have other values; blue and white tend to have higher current requirements than green or red, and LEDs sold as 'ultrabright' have still higher needs.

Often the value you get from the equation doesn't map to any common value of resistor, in which case you simply round up to the nearest resistor value you have to hand. In this example, a 68Ω resistor protects the LED perfectly with only a small loss of brightness; rounding down to a 47Ω resistor may also work, but with a wide enough tolerance may allow enough current to flow to damage the LED over a long enough time.

Note, however, that the BBC micro:bit's edge connector is not designed for high-current devices. Although your chosen LED may have a forward current of 25mA, the BBC micro:bit's edge connector is only capable of providing around 0.5mA before the voltage begins to drop below its usual 3V. For individual LEDs, this shouldn't matter. You may find your LED is dimmer than you would expect, but it won't do any damage. For other devices, such as motors or long strips of multiple LEDs, a given pin on the edge connector will switch off to protect the BBC micro:bit long before providing enough current to run the device. In these cases, you will need an external power supply and a transistor or other controlling component to activate and deactivate the high-current device.

If you have no documentation for your LEDs—if they were salvaged from old electronics, for example—then always err on the side of caution. Try higher-value resistors first, and if the LED seems too dim, step down until you find a suitable value. Remember, you can always replace the resistor, but when the LED has blown, it's gone for good!

Put your circuit together using the following instructions so it matches Figure 10-8, as always making sure that if you're using crocodile clips they are attached to the BBC micro:bit securely and not accidentally bridging any of the smaller pins and beginning with the BBC micro:bit switched off.

1. Take the LED between your fingers and look at its metal legs. The longer leg is called the *anode*, and the shorter one is called the *cathode*. Attach a crocodile clip wire to the longer leg (anode), and then connect the other end of the wire to Pin 1 of the BBC micro:bit using either a crocodile clip or a banana plug.

2. Attach another crocodile clip wire to the shorter leg (cathode) of the LED, and then attach the other end of the wire to one leg of your current-limiting resistor. The resistor works either way around, so don't worry about which direction the coloured bands are facing.

3. Attach a final crocodile clip wire to the other leg of the resistor, and then connect the other end to the GND pin on the BBC micro:bit using a crocodile clip or banana plug to complete the circuit.

Next, you need to write a program to make the LED do something so you know it's working. A good way to start is by simply blinking the LED on and off every second, which lets you know everything is wired up properly and that the BBC micro:bit is running your program correctly.

Load the Python editor in your web browser and enter the following program code:

```
from microbit import *
while True:
    pin1.write_digital(1)
    sleep(1000)
    pin1.write_digital(0)
    sleep(1000)
```

Remember that indented lines in Python should always have four spaces for each indent (see Chapter 7 for more details); in this program, everything after the line `while True:` should be indented.

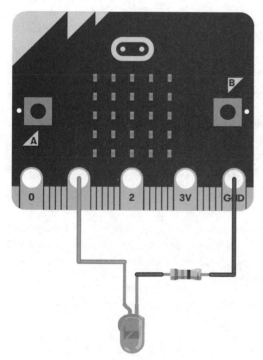

FIGURE 10-8: A simple LED circuit

After double-checking all your wiring, connect the BBC micro:bit to your computer with a micro-USB cable, click on the Download button, and drag the hex file to your BBC MICROBIT drive to flash the program. The LED should start flashing on and off once every second as the program runs.

If the LED doesn't light up, check your wiring. Common problems include having the LED the wrong way around or connecting the crocodile clips to Pin 0 or Pin 2 rather than Pin 1.

Step by step, this program does the following:

```
from microbit import *
```

This line, as with all Python programs written for the BBC micro:bit, imports the code necessary to address the BBC micro:bit. Without it, the program doesn't work.

```
while True:
```

As before, this line puts the BBC micro:bit into an infinite loop. So long as True is true—which it always is—the code below this line runs continuously. Without this line, the LED flashes once and then stops as the program reaches its end.

```
    pin1.write_digital(1)
```

You're addressing Pin 1 this time, rather than Pin 0, but the basic format is the same. This line starts with the pin you want to control followed by a full stop and then exactly what you want to do to it. Rather than reading from it, this time you're outputting a signal—which is known as *writing* to the pin, in contrast with reading from the pin. A value of 1 in the brackets turns the pin on, making electricity flow through the LED and light it up.

```
    sleep(1000)
```

The BBC micro:bit's processor runs so quickly that it is capable of turning the LED on and off faster than the eye can see. This line slows it down, telling the BBC micro:bit to wait 1,000 milliseconds (one second) before moving on to the next line.

```
    pin1.write_digital(0)
```

The only difference between this line and the one that turned the LED on is the value written to the pin: a 1 turned it on, and this 0 turns it off again. The `write_digital` instruction can only write these two binary values; later in this chapter you see how to use a different instruction to fake a signal between 0 and 1.

```
sleep(1000)
```

Finally, this line tells the BBC micro:bit to sleep for another second before starting again at the top of the infinite loop. Both delays are needed; otherwise, the LED turns on so quickly after turning off you never have a chance to see the difference.

To experiment with the program, try changing the delays. Higher numbers lengthen the time between instructions, while lower numbers shorten the delay. Try making the two delays different values to have the LED stay on for longer than it stays off or vice versa—or even lengthen the program with multiple `write_digital` and `sleep` instructions to have it flash in a pattern.

Fading an LED via PWM

In the last circuit, you saw how the `write_digital` instruction can be used to turn an LED on and off. The BBC micro:bit can do more than simply turn its pins on and off, however. It can make them seem to fade smoothly between the two or sit at any value between fully on and fully off.

The BBC micro:bit does this using a technique called *pulse-width modulation* (PWM). While it's not actually capable of doing anything other than turning its pins fully on or fully off, it can turn them on and off so quickly that they appear to be somewhere between fully on and fully off—literally *pulsing* the signal. By changing, or *modulating*, the amount of time the pin spends in the on and off state during each pulse cycle—known as the *duty cycle*—you can change the brightness of an LED or the speed of a motor.

For this circuit, you can use the same layout as in the previous circuit with an LED and a suitable current-limiting resistor connected between Pin 1 and the GND pin of the BBC micro:bit. The program, however, is different.

Load the Python editor in your web browser and type in the following program:

```
from microbit import *
pin1.set_analog_period(1)
while True:
    for brightnessValue in range(0,1024,1):
        pin1.write_analog(brightnessValue)
        sleep(1)
    for brightnessValue in range(1023,-1,-1):
        pin1.write_analog(brightnessValue)
        sleep(1)
```

Double-check your wiring, and then connect the BBC micro:bit to your computer with the micro-USB cable. Click the Download button to download your program, and then drag the file to the MICROBIT drive to install and run it. After a few seconds, the LED should begin smoothly pulsing from dark to bright and back again.

Step by step, the program does the following:

```
from microbit import *
```

As always, this line imports the instructions you need to communicate with the BBC micro:bit.

```
while True:
```

This begins the infinite loop, so the program doesn't finish when it comes to the final instruction.

```
pin1.set_analog_period(1)
```

This line tells the BBC micro:bit to set the *period* of the pulse-width modulation signal to one millisecond. The period controls how long each cycle of the pulse-width modulation cycle lasts: the longer the period, the longer the LED spends in its on and off states. Too long, and you see a visible flicker, while a period of one millisecond should give the appearance of the LED being constantly lit but at different brightnesses.

```
    for brightnessValue in range(0,1024,1):
```

This instruction begins a different type of loop. Rather than infinite, this `for` loop has a defined stopping point: when the output of the `range` instruction ends. The `range` instruction itself takes three inputs: the number it should start at, the number it should stop before, and the amount it should count by each time. In this case, it's starting at 0, stopping just before reaching 1024, and each count adds 1 to the number; in other words, it puts values from 0 to 1023 in the variable `brightnessValue`.

```
        pin1.write_analog(brightnessValue)
```

This line uses the `write_analog` instruction to write a value to Pin 1, but where the `write_digital` instruction only accepts a 0 or a 1, the `write_analog` instruction accepts any value from 0 to 1023. Rather than writing the value directly, though, this line uses the value stored in the `brightnessValue` variable.

```
    sleep(1)
```

As always, the BBC micro:bit runs too quickly for the fading effect to be easily visible; this line tells the BBC micro:bit to wait one millisecond between each iteration of the loop making the transition more easily visible.

```
for brightnessValue in range(1023,-1,-1):
```

The only difference between this line and the previous `for` loop is in the values given to the `range` instruction. This time, the count starts at 1023, where the previous count ended, stops just before –1, and counts by –1 each time. In other words, this line is the exact opposite of the previous loop: it counts backward from 1023 rather than forward to 1023, turning the LED from light to dark rather than dark to light.

```
    pin1.write_analog(brightnessValue)
```

As before, you need to write the value to Pin 1 to actually change the brightness of the LED, which this `write_analog` instruction does.

```
    sleep(1)
```

Finally, a last `sleep` delays the process long enough to see its effect. As with the flashing example, changing the `sleep` delays changes the speed of the effect; experiment with different values to see how they affect the smoothness of the fade.

This same method can control hardware that requires a PWM signal, such as stepper motors. To do so, you may need to change the period of the pulse; a short period suitable for dimming an LED may not be appropriate for a stepper or servo motor. More information on changing the period of a PWM pulse is available in Appendix C, 'Python Recipes'.

Reading an Analogue Input

A traditional switch or a button is a digital input; it's either pressed or it's not, with nothing in between. Not all inputs are digital, however. Some are capable of offering a range of values between fully on and fully off, and these are called *analogue inputs*.

All three of the large input pins of the BBC micro:bit support reading analogue inputs thanks to the board's analogue to digital converter (ADC), which takes the analogue inputs and transforms them into a digital format the BBC micro:bit can understand. For this project, you use Pin 2 of the BBC micro:bit.

To build this circuit, you need the following:

- Wires with crocodile clips or banana plugs
- A potentiometer

A potentiometer is, in effect, a form of variable. When at its minimum setting, a potentiometer is almost completely without resistance—like a plain piece of wire; at its maximum setting, a potentiometer offers its full rated resistance—10KΩ or 100KΩ for the most commonly used potentiometers in electronics kits. At its halfway point, it offers half its rated resistance—5KΩ or 50KΩ. Some have a sliding mechanism, but the most commonly used versions have a knob you can turn; both operate in the same manner.

Put your circuit together using the following instructions so it matches Figure 10-9, as always making sure that if you're using crocodile clips, they are attached to the BBC micro:bit securely and not accidentally bridging any of the smaller pins and beginning with the BBC micro:bit switched off.

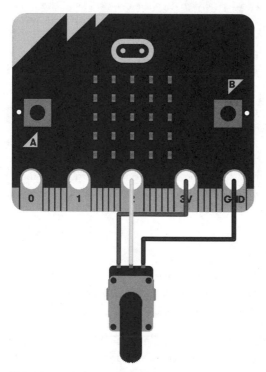

FIGURE 10-9: A potentiometer circuit

1. Holding the potentiometer so that its knob is facing toward you, attach a crocodile clip wire to the leftmost of its three legs, and then attach the other end to the 3V pin of the BBC micro:bit with a crocodile clip or banana plug.

2. Attach a second crocodile clip to the rightmost of the potentiometer's three legs, and then attach the other end to the GND pin of the BBC micro:bit with a crocodile clip or banana plug.

3. Attach a final crocodile clip wire to the middle leg of the potentiometer, and then connect the other end to Pin 2 of the BBC micro:bit using a crocodile clip or banana plug to complete the circuit.

The potentiometer circuit works by providing the potentiometer with two input signals: one at the maximum voltage we want to use, from the 3V pin, and another at the minimum voltage, from the GND pin. As the knob is turned or slider slid, the potentiometer acts as a voltage divider and outputs from the middle pin a voltage somewhere between the two inputs depending on how far the knob is turned.

To see the potentiometer in action, you need to tell the BBC micro:bit to read an analogue signal; otherwise, you just see the signal from the potentiometer jump from 0 to 1 partway through turning the knob. Load the Python Editor in your web browser and enter the following program code:

```
from microbit import *
while True:
    potValue = pin2.read_analog()
    display.scroll(str(potValue))
```

Double-check your wiring, and then connect the BBC micro:bit to your computer with the micro-USB cable. Click the Download button to download your program, and then drag the file to the MICROBIT drive to install and run it. After a few seconds, a series of numbers should begin scrolling across the screen. Try turning the knob or sliding the slider of the potentiometer and see what happens to the numbers.

Step by step, the program does the following:

```
from microbit import *
```

A Python program can't have access to the BBC micro:bit's pins and other hardware without this line, so it's no surprise to find it here at the start of this program.

```
while True:
```

This program again uses an infinite loop to prevent it from stopping as soon as a single value has been read.

```
potValue = pin2.read_analog()
```

This creates a variable called potValue and then takes an analogue reading from Pin 2 and stores it there. The output of a read_analog instruction is always a number between 0 and 1023, where 0 represents fully off or 0V and 1023 represents fully on or 3V.

```
display.scroll(str(potValue))
```

The value read in the line above is now printed to the display using the display.scroll instruction. Because the reading is an integer data type, however, it has to first be converted to a string ready for printing. The str instruction handles this.

Unlike earlier programs, there's no need for a delay; printing the value to the BBC micro:bit's display pauses the program until it's complete, introducing a natural delay and giving you a chance to read the numbers.

With these circuits under your belt, it's time to flex your skills. Try connecting the button, the LED, and the potentiometer to the BBC micro:bit at the same time, reading from the potentiometer and using that information to control the brightness of the LED and the button to turn it on and off. Once you've mastered that, you're well on your way to building your own circuits which can extend the BBC micro:bit's capabilities in all kinds of ways.

Chapter 11

Extending the BBC micro:bit

In this chapter

- Extending the BBC micro:bit with breakout boards
- The BBC micro:bit and robotics: off-the-shelf solutions to building a BBC micro:bit-powered robot
- Other notable BBC micro:bit add-ons

THE BBC MICRO:BIT is a powerful device in its own right, but it really comes into its own when it's paired with hardware that expands its capabilities. From simple breakout boards which provide easy access to the 20 small pins spaced along its edge connector to kits for building your own autonomous robots, this chapter introduces some of the accessories available for building more complex projects.

Only use add-on boards and devices which are sold as being compatible with the BBC micro:bit. **WARNING** Microcontrollers use a range of different voltages, and hardware made for a microcontroller running at 5V will at best fail to work properly with the 3V BBC micro:bit or at worst could damage your BBC micro:bit.

Extending via Breakout Boards

The large pins on the BBC micro:bit's edge connector are enough to build small projects, but eventually you'll find yourself needing more. A breakout board provides easy access to the smaller pins on the edge connector, multiplying the amount of hardware you can add while providing access to the more advanced bus types.

The available breakout boards vary in pricing, in features, and in how they make the pins available. A selection of different boards is offered next as a representative example.

Kitronik Edge Connector Breakout Board

The Kitronik Edge Connector Breakout Board (see Figure 11-1), created in the company's role as one of the founding partners of the BBC micro:bit project, is designed to make it easier to use with BBC micro:bit with external hardware. Every pin on the BBC micro:bit—both the large pins and the small pins—is broken out into easily-accessible 2.54mm pin headers, which you can use with 2.54mm female-to-male jumper wires to connect the BBC micro:bit to a breadboard.

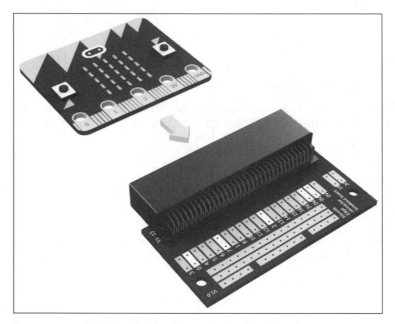

FIGURE 11-1: The Kitronik Edge Connector Breakout Board

Kitronik's design is notable for making it as simple as possible to connect hardware to the BBC micro:bit's additional pins. Every major pin from the BBC micro:bit's edge connector is broken out to two male 2.54mm pin headers so you can make more than one connection, while there's an unpopulated prototyping area at the bottom of the board which provides additional 3V and GND connections plus three linked groups for your own use.

The Edge Connector Breakout Board is available in two versions: a standard version and a prebuilt version. The standard version comes with the edge connector itself disconnected from the breakout board and the pin headers unpopulated, and it requires over a hundred solder connections before it can be used. The prebuilt version, by contrast, comes with the edge connector and most pin headers already soldered into place; only Pin 19 and Pin 20, used for the I²C bus, are left unpopulated for you to solder yourself, while the prototyping area is left bare.

You can find more information on the Kitronik Edge Connector Breakout Board at kitronik.co.uk.

ScienceScope Micro:bit Breakout Board

Designed specifically for classroom use but also well suited to home users with younger children, the Micro:bit Breakout Board from ScienceScope (see Figure 11-2) uses child-friendly 4mm banana plug connectors. These are the same connectors as you'll find on the micro:bit's large pins, meaning it's possible to take your existing banana plug cables and reuse them.

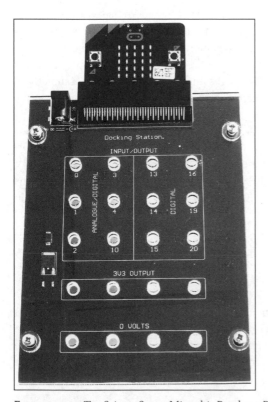

FIGURE 11-2: The ScienceScope Micro:bit Breakout Board

While some breakout boards provide access to all or most of the smaller pins on the BBC micro:bit, the ScienceScope version concentrates only on the general-purpose input-output (GPIO) pins. Besides breaking out the large Pins 0 through 2, ScienceScope's board provides easy access to the digital and analogue-to-digital Pins 3, 4, and 10, and the digital-only Pins 13, 14, 15, 16, 19, and 20.

The ScienceScope board also includes four linked outputs for 3.3V power and GND connectivity, along with a dedicated power supply input for powering the micro:bit and the power connectors. This gives the board greater scope for powering higher-current devices, such as ultra-bright LEDs, than alternatives which rely purely on the micro:bit's internal power supply.

You can find more information on the ScienceScope Micro:bit Breakout Board at sciencescope.uk.

Proto-Pic Bread:Bit

The Proto-Pic Bread:Bit breakout board (see Figure 11-3) is designed for use with a solderless breadboard, which is not included in the package. Rather than connecting wires directly to the breakout board, the Bread:Bit is inserted into a solderless breadboard and connections made through that instead. For anyone looking to build more complicated projects, this makes for a more stable build with fewer trailing wires and can even be soldered directly to stripboard to build a permanent add-on circuit from which the micro:bit can be easily removed and reconnected.

The Bread:Bit provides connections for all 25 of the pins on the micro:bit, making it suitable for any project. Each pin is given a single connection on the breadboard, including the 3V and GND pins. The board is provided with the edge connector soldered into place, but the 2.54mm male pin headers left off; for an extra fee, it's possible to buy a version with the pin headers already soldered into place.

You can find more information on the Proto-Pic Bread:Bit at proto-pic.co.uk.

Proto-Pic Exhi:Bit

The second prototyping board design from Proto-Pic, the Exhi:Bit (see Figure 11-4), is designed to be fully comprehensive. Where the Bread:Bit is designed to be used with an optional solderless breadboard, the Exhi:Bit comes with its own, which can be slotted into the middle of the board. For more permanent projects, the central area also includes a stripboard-like prototyping area onto which you can solder your own components.

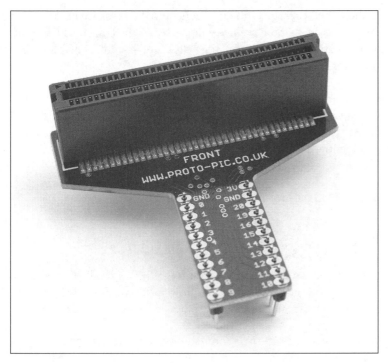

FIGURE 11-3: The Proto-Pic Bread:Bit

FIGURE 11-4: The Proto-Pic Exhi:Bit

The feature that makes the Exhi:Bit really stand out, though, is the presence of large pins running across three of its four sides. These are near-duplicates of the large pins found on the micro:bit itself, and compatible with both crocodile clips and 4mm banana plugs, but every pin is available rather than just the default five.

Proto-Pic has also designed a series of daughterboards which can be connected to female pin headers on the Exhi:Bit, expanding its capabilities, as well as a do-it-yourself board onto which you can easily solder your own components. There are also three note-taking areas, designed primarily for differentiating multiple units in a classroom setting, and a dedicated 5-9V power input.

You can find more information on the Proto-Pic Exhi:Bit at `proto-pic.co.uk`.

Robotics and the BBC micro:bit

Despite its small size, the micro:bit is a powerful tool. Its internal sensors and relatively large number of programmable pins make it ideal for robotics projects. It's small and light enough not to weigh down whatever you build, and its low power draw means longer battery life.

You can build your own robot using off-the-shelf components, but a number of companies have begun offering robot kits. Simply pick up a kit, assemble the robot, program the micro:bit, and slot it home to complete the build.

A selection of different robot kits is offered next as a representative sample.

Kitronik Line-Following Buggy

A relatively simple two-wheeled robot, Kitronik's Line-Following Buggy (see Figure 11-5) adds a pair of extra sensors to the ones on the micro:bit itself: light-dependent resistors (LDRs). Using the LDRs and LED lights that shine underneath the buggy, the robot is capable of following a black line drawn on the floor without the need for a remote control.

The kit includes everything you need to build the robot, including a motor driver board which allows the micro:bit to control the twin motors and a battery pack which holds four AA batteries to power both the motors and the micro:bit itself. Two sample programs are also provided: one which follows a line drawn on the floor, and the other which uses the micro:bit's accelerometer to spin the robot around when it bumps into obstacles.

More information on the Kitronik Line-Following Buggy kit can be found on `kitronik.co.uk`.

FIGURE 11-5: The Kitronik Line-Following Buggy

Kitronik Motor Driver Board

Provided as part of the Line-Following Buggy kit, Kitronik's Motor Driver Board (see Figure 11-6) is also sold on its own for those looking to build their own custom robotics projects. A BBC micro:bit on its own can't provide enough power to turn a motor on and off; the Motor Driver Board fixes that, providing connections to drive two small DC motors.

The Motor Driver Board also includes 2.54mm pin headers for accessing the BBC micro:bit's pins—though you need to solder the pins on yourself if you want to use them—as well as screw terminal connectors for the motors, Pin 1 and Pin 2, Pin 5 and Pin 11 (equivalent to pressing Button A and Button B on BBC micro:bit itself), and a 4.5-6V power input for three or four AA batteries.

You can find more information on the Kitronik Motor Driver Board at `kitronik.co.uk`.

FIGURE 11-6: The Kitronik Motor Driver Board

Technology Will Save Us Micro:Bot

Designed by Technology Will Save Us, one of the partner companies in the BBC micro:bit project, the Micro:Bot kits (see Figure 11-7) are primarily designed to be affordable. Rather than using a plastic or metal chassis, as with many robot kits, the body of the robot is made from the cardboard box in which the robot is shipped.

Technology Will Save Us has developed three projects around the Micro:Bot: ArtBot, RoomBot, and GolfBot. You can use a single kit to build all three—plus a few common household items, such as a pen, some tinfoil, and a kitchen sponge. ArtBot turns the Micro:Bot into a basic line-drawing turtle by poking a pen through the cardboard body; RoomBot uses tinfoil to make a collision sensor; and GolfBot adds a putting green to the front to make a golf trainer that relocates itself on a successful put.

FIGURE 11-7: The Technology Will Save Us Micro:Bot

You can find more information on the Technology Will Save Us Micro:Bot kit at techwillsaveus.com.

4tronix Bit:Bot

The principle behind the 4tronix Bit:Bot (see Figure 11-8) is simple: to provide as many features as possible on the body of a two-wheeled BBC micro:bit robot chassis. The kit includes the same features as the Kitronik Line-Following Buggy, with two geared motors and light-dependent resistors for line-following functionality, plus a range of extras: two sets of six programmable multi-colour LEDs at each side of the body, two light sensors which allow the robot to follow a light source such as a torch, and a buzzer for making simple sounds.

The Bit:Bot can also be expanded to add additional features. An expansion port on the main part of the body allows additional programmable LED boards to be added, while a front expansion port provides connectivity for an ultrasonic distance sensor which can measure how far away the robot is from any obstacles it may find in its path. Code snippets are available on the official website, along with sample Python programs for controlling the motors, LEDs, and sensors and making the robot act as a line follower.

You can find more information on the 4tronix Bit:Bot at `4tronix.co.uk`.

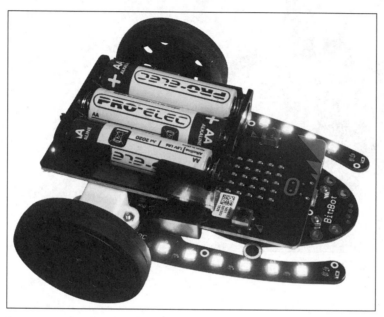

FIGURE 11-8: The 4tronix Bit:Bot

BinaryBots

Designed to appeal to younger children, the BinaryBot kits (see Figure 11-9) are nothing if not colourful. The body of each robot is built from cleverly folded cardboard, onto which colourful stickers can be applied to make either a traditional-looking bipedal robot or a saucer-shaped UFO.

Unlike the other robot kits in this chapter, however, the BinaryBots aren't true robots; they can't move. Each kit includes a speaker, temperature sensor, motion sensor, light sensor, and sample programs for reading the input of each. While the BBC micro:bit can then use these devices to make noises and flash its lights, though, a lack of motors means they're entirely stationary—which in a busy household or classroom full of younger pupils may be a positive rather than a negative!

You can find more information on the BinaryBots kits at `binarybots.co.uk`.

FIGURE 11-9: The BinaryBots

Other BBC micro:bit Add-Ons

The add-ons available for the BBC micro:bit are not limited to robots and breakout boards. Developers, engineers, and students all around the world have designed their own add-ons which use the edge connector. Some of the add-ons let you play games, while others make it easier to build portable and wearable projects around the BBC micro:bit.

A selection of these is offered next as a taste of what's out there.

Kitronik Mi:Power

The Mi:Power board from Kitronik (see Figure 11-10) is designed to simplify portable and wearable projects based around the BBC micro:bit by doing away with a separate battery holder. Designed to match the footprint of the BBC micro:bit, the Mi:Power board connects via bolts and spacers to Pin 0, 3V, and GND on the BBC micro:bit and allows it to draw its power from an on-board 3V coin-cell battery—protected, thankfully, from being removed and potentially swallowed by younger children by being located on the inside of the micro:bit-Mi:Power 'sandwich', and accessible only by undoing the three screws holding the boards together.

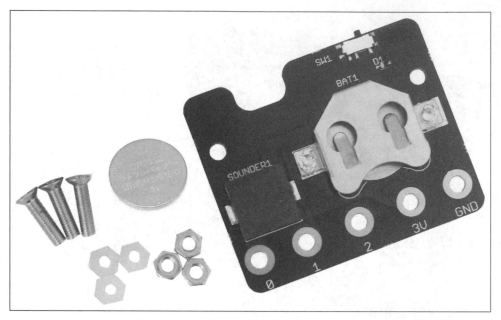

FIGURE 11-10: The Kitronik Mi:Power

Besides providing power to the BBC micro:bit, the Mi:Power board includes a buzzer connected to Pin 0 and a dedicated on-off switch that allows the BBC micro:bit and connected hardware to be powered off without having to disconnect wires. For those with access to a laser cutter, as part of a home workshop, at a school or university, or at a local hackspace or maker space, Kitronik provides files for cutting add-on accessories, including a watch mount, plant spike, lanyard attachment, belt mount, and even small robot-like character.

You can find more information on the Mi:Power at `kitronik.co.uk`.

Proto-Pic Micro:Pixel Board

The 5×5 LED display on the front of the BBC micro:bit is enough to display a surprising amount of information, but eventually it's possible to tire of its red light. Proto-Pic's answer to this is the Micro:Pixel board (see Figure 11-11), which plugs into the edge connector and provides a 4×8 matrix of individually-programmable full-colour LEDs—allowing you to reproduce almost any colour you can imagine.

The larger pins of the BBC micro:bit are duplicated on the edge of the Micro:Pixel board, though the smaller pins are not and Pin 0 is used to communicate with the LEDs themselves. The board also comes with a warning: the LEDs at full brightness can draw more power than the BBC micro:bit is capable of providing, making it theoretically possible to damage the BBC micro:bit by setting the LEDs too high—something to bear in mind while experimenting with the board.

You can find more information on the Micro:Pixel Board at `proto-pic.co.uk`.

FIGURE 11-11: The Proto-Pic Micro:Pixel Board

Proto-Pic Simon:Says Board

The Proto-Pic Simon:Says board (see Figure 11-12) serves three main purposes: it's a great demonstration of how capacitive touch can turn parts of a printed circuit board into 'buttons'; it's a re-creation of a beloved electronic game first invented by Ralph Baer, Howard Morrison, and Lenny Cope in 1978; and it can act as a four-button input with LEDs and a buzzer for your own projects.

Just like the original, Proto-Pic's version of Simon is primarily a game: the sample code provided lights up the LEDs and sounds the buzzer in a pattern you have to copy, getting faster and more difficult each time. When you're tired of losing to the BBC micro:bit, though, it's a surprisingly powerful expansion board in its own right, with four programmable colour LEDs, four capacitive-touch buttons, and a buzzer, all of which can be integrated into your own programs with ease.

You can find more information on the Simon:Says board at `proto-pic.co.uk`.

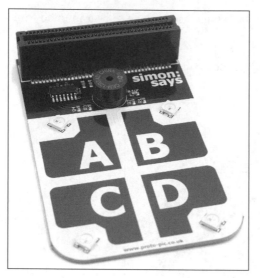

FIGURE 11-12: The Proto-Pic Simon:Says Board

4tronix Bit:2:Pi Board

The Raspberry Pi and the BBC micro:bit are great companions, as proven in Chapter 9, 'The BBC micro:bit and the Raspberry Pi'. The 4tronix Bit:2:Pi board (see Figure 11-13) takes things even a step further by allowing you to connect selected Raspberry Pi add-ons directly to the BBC micro:bit—including the various Hardware Attached on Top (HAT) boards available for the Pi.

FIGURE 11-13: The 4tronix Bit:2:Pi Board

Not every Raspberry Pi add-on works with the BBC micro:bit, but many can be convinced to operate if you're willing to spend the time to write the necessary code. Examples of add-ons you can easily use with the Bit:2:Pi include those based around programmable LEDs, buttons, buzzers, and motor controllers. Those that won't work include display screens and cameras.

You can find more information on the Bit:2:Pi board, including a list of Raspberry Pi add-ons that are known to be compatible, at `4tronix.co.uk`.

Kitronik Mi:Pro Protector and Mi:Power Cases

Although the BBC micro:bit is designed to be robust, if you're taking it out and about it's a good idea to protect it from accidental damage. The Kitronik Mi:Pro Protector case (see Figure 11-14) does exactly that, covering the front and back of the BBC micro:bit while still providing access to the micro-USB connector, edge connector, and buttons.

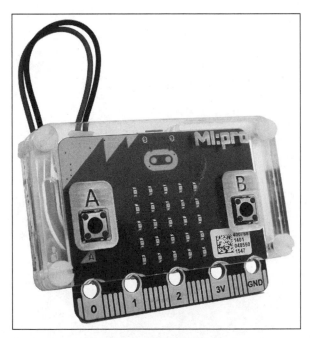

FIGURE 11-14: The Kitronik Mi:Pro Protector

For powering the BBC micro:bit on the move, an optional battery holder for two AAA batteries can be screwed to the back of the Mi:Pro Protector. Alternatively, a version for use with the Mi:Power battery board is available, with both versions being produced in clear, green, orange, and blue coloured plastic.

You can find more information on the Mi:Pro Protector and MiPower cases at `kitronik.co.uk`.

Chapter 12

The Wearable BBC micro:bit

In this chapter

- A look at what makes the BBC micro:bit so well suited to wearable and embedded projects
- An explanation of conductive thread
- The Rain-Sensing Hat: A semi-practical project for you to build

WHILE THE WORLD'S first computers were literally the size of a room—or, in some cases, several rooms—the march of technology has given us devices of greater processing power which we can fit in our pockets. As soon as electronics became small enough, people started to find ways to carry them around at all times. Pocket calculators led to digital watches, which eventually gave rise to the modern smartwatch (see Figure 12-1) and related devices like fitness trackers and location monitors.

Although most people's main experience of wearable electronics is for tasks like schedule notification and fitness monitoring, the technology is used in a variety of fields. Hardware is being developed which aims to monitor the wearer's health, with tiny pill-like sensors which can be swallowed to monitor internal health, and there are even devices which can actively improve conditions such as Parkinson's Disease.

As electronics become ever-smaller and more powerful, it's likely that the field of wearable electronics will steadily grow—and the BBC micro:bit is a great platform to begin your own experiments in the field.

FIGURE 12-1: A smartwatch-style wearable

Advantages of the Wearable BBC micro:bit

The BBC micro:bit has a range of features which make it extremely well suited to wearable projects, not the least of which is its size. Measuring around half the size of a standard credit card, the BBC micro:bit is small and light enough to be integrated into clothing or worn directly on the body without causing discomfort.

Another feature of the BBC micro:bit which makes it easier to build practical wearable electronics project is its low power draw. A small laptop computer can run from its battery for around four to eight hours; a modern smartwatch can go for between one and four days between charges; a BBC micro:bit running from a pair of AA batteries (see Figure 12-2), however, can gather information from its sensors for over a month without difficulty.

The BBC micro:bit's built-in display, sensors, and buttons make it an all-in-one device, too. Traditionally, a microcontroller is provided without these accessories, meaning you need to add them on yourself to build your wearable project. Naturally, the input-output pins on the BBC micro:bit mean that you can add extra hardware if your project requires it.

FIGURE 12-2: A BBC micro:bit running from batteries

There's a key extra feature that makes the BBC micro:bit particularly well suited to wearable projects, though, and it's one not commonly seen in microcontrollers: support for conductive thread, which can be wound around the holes in the larger input-output pins to quickly make a solid electrical contact for fabric-based flexible, soft, and wearable circuits.

Conductive Thread

For a circuit to work, electricity has to be able to flow along *conductors*—the opposite of *insulators*—which are typically copper wires or copper tracks on a printed circuit board. The copper-coloured input-output pins on the BBC micro:bit are conductors, while underneath the BBC micro:bit's surface are more copper conductors called *tracks* which serve to connect its components together.

Copper wire is an excellent conductor, but it has disadvantages. Chief among these is its weakness in the face of constant movement: as a wire bends backward and forward, the metal begins to weaken—a process known as *metal fatigue*—and eventually breaks. Once the wire has broken, it can no longer act as a conductor, and the circuit fails. The same is true of solder joints connecting wires to other components. Pressure on these solder joints can cause the solder or the copper to which it is connected to weaken and break, again causing the circuit to fail.

Put together, these issues are a real problem for wearable projects. This is where conductive thread comes in. As the name suggests, conductive thread is a sewable thread—similar to a thick cotton or woven nylon—which acts as a conductor, replacing the copper wire in circuits with a material that is flexible and soft to the touch (see Figure 12-3).

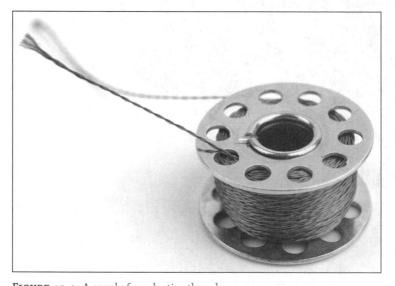

FIGURE 12-3: A spool of conductive thread

Conductive thread is typically made of extremely fine strands of steel, sometimes with silver plating, woven in the same way as cotton to make a thick and strong thread (see Figure 12-4). As a rule of thumb, the thicker the thread, the more conductive it is and the more power it can supply to components. A thin thread may provide enough power for a small LED or a button, while a thicker thread—or multiple strands of thinner thread woven together— could provide enough power for a small display or motor.

While copper wire is usually covered in plastic as an insulator, conductive thread is not. If you touch two conductive threads together, they act as a single conductor. As a result, you have to be careful when making a circuit with conductive thread. If you accidentally cross two separate sections of conductive thread, they create a short-circuit which at best prevents your project from working properly and at worst can damage components up to and including the BBC micro:bit itself.

FIGURE 12-4: A length of conductive thread under a microscope

The lack of insulator on conductive thread makes it easy to work with, however, and even allows you to build touch or moisture sensors without a single piece of extra hardware, as you'll see later in this chapter.

Using Conductive Thread

Conductive thread is easy to work with, but you need a few extra pieces of hardware to get started. Besides the thread itself, which is sold at most hobbyist electronics shops and websites under a variety of brand names, you need the following:

- **A sewing needle**—When using conductive thread to add a circuit to fabric, you need a sewing needle to pull it through. Make sure to consider the size of your thread when choosing a needle. Too small a needle, and you cannot insert your conductive thread into its eye; too large a needle, and you're accidentally pulling the end of the thread through the hole you just made.

- **Fabric**—A conductive thread circuit needs an insulator, just like a printed circuit board, and this is typically fabric. Usually, this is whatever you are looking to build your wearable project on; gloves, a hat, a coat, and shoes are all suitable insulators. The thicker the material, though, and the harder it is to sew. Thick leather is considerably more difficult to work with than thin cotton.

- **A needle threader**—Although not strictly necessary, a needle threader makes it easier to thread the needle with your conductive thread. This is particularly important when using thicker thread, which can fray at the end making it difficult to fit through the needle without help. Needle threaders are readily available from craft and fabric shops, often in a kit with needles.

- **A thimble**—As with the needle threader, you don't strictly need a thimble, but if you're working with tougher fabric or are inexperienced with sewing, it makes life a lot easier. Usually made of metal or ceramic, the thimble goes over your thumb or finger and prevents you from accidentally pricking yourself with the needle as it passes through the fabric.

- **Scissors**—You need to be able to cut lengths of thread to size, and for that you need a small but sharp pair of scissors. Try to keep a dedicated pair for this task. Conductive thread is made of metal and can leave blunt or dented sections in scissor blades, so having a pair you only use for cutting conductive thread stops you from having issues using them to cut anything else afterward.

If you've sewn with traditional cotton thread before, using conductive thread should be immediately familiar, and you can skip ahead to the practical project later in this chapter. If not, follow the guide next to get started.

Start by cutting a length of conductive thread from the spool with your scissors. Always err on the side of having excess thread left over on your length after you've finished, and remember that you may be sewing in zig-zag patterns, which further increases the amount of thread you use compared to the distance it covers on your fabric.

> **TIP** If you're just practising and don't want to waste your conductive thread, buy a spool of cheap cotton thread of around the same thickness. Although it doesn't conduct electricity, it allows you to practice sewing without wasting conductive thread.

Take the opposite end of your length of thread to the one you just cut and tie a knot in it. This knot stops the thread from being pulled through the hole the needle makes as you begin sewing, so it needs to be reasonably large; try looping the thread through the knot several times over before tightening it to increase its size (see Figure 12-5).

Take the end of the conductive thread you just cut and pass it through the eye of the needle, pulling a short length through to hold (see Figure 12-6). Conductive thread frays quickly, so if you find it difficult to thread through the needle, consider making a fresh cut, folding the thread and passing the folded portion through the needle instead, or using a needle threader to make the process easier.

You're now free to begin sewing. Placing a thimble on the finger or thumb on the other side of the fabric, push the needle carefully through. When the needle is around halfway through, switch to the other side of the fabric and begin pulling until the needle is all the way through and is pulling the conductive thread through the fabric.

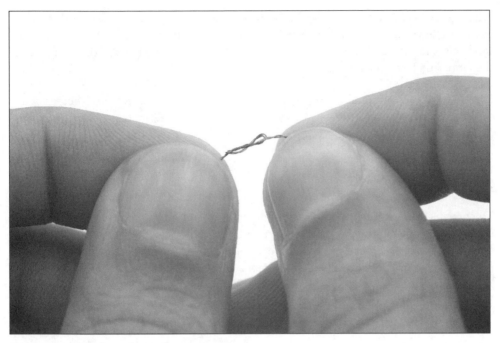

FIGURE 12-5: Knotting the conductive thread

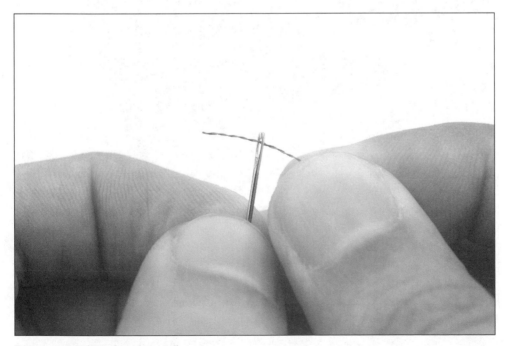

FIGURE 12-6: Threading the needle

When gently pulling the rest of the thread through, take care not to let the thread come out of the eye of the needle. If it does, just rethread it and continue sewing. Once the knot has reached the fabric, the thread will stop; don't pull it any harder, or you'll pull the knot through the fabric and have to start again.

Exactly how you continue to sew depends on the fabric you're using and whether you're sewing to create a strong join—such as when linking two separate pieces of fabric —or simply to create a conductive path for your circuit. As a rule of thumb, the more times the needle passes through the fabric, the stronger the join. A length of thread which sits on top of the fabric and only passes through at two points is weaker than one which constantly zig-zags backward and forward.

To make a connection to one of the BBC micro:bit's input-output pins, simply wrap a free end of thread around the pin, passing it through the hole and back again (see Figure 12-7). Make sure to wind it around a few times, to make a secure and electrically-sound connection. If you're attaching the BBC micro:bit to the fabric at this point, you can do this by passing the needle through the fabric and the hole of the BBC micro:bit's input-output pin, and then looping it back to the fabric at the bottom of the pin before repeating the process four or five times. This gives you a solid and secure connection.

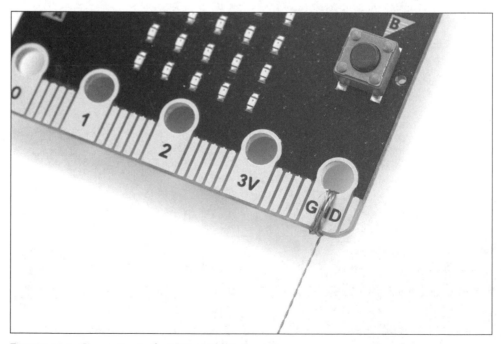

FIGURE 12-7: Connecting to the BBC micro:bit

The Rain-Sensing Hat

As a demonstration of how to build a simple wearable using the BBC micro:bit, a rain-sensing hat is a fantastic project. As an actual practical device, it's somewhat less useful. If you're outside and it's raining, you know about it whether or not you're wearing a 'smart' hat!

To follow this project, you need the following materials:

- **A BBC micro:bit**—The BBC micro:bit forms the brains of the wearable, actually detecting when the rain begins to fall, as well as providing the display that warns the wearer to seek shelter. You also need a micro-USB cable to program the BBC micro:bit from your computer.

- **A battery pack**—A wearable that keeps you tethered to your computer through a USB cable isn't much of a wearable, so you need a battery pack. The one provided as part of the micro:bit Go bundle is perfect, or you can reduce the weight of the hat by opting for a third-party pack which uses smaller batteries. You also need fresh alkaline batteries, rather than part-used or rechargeable batteries.

- **Conductive thread**—You use the conductive thread to build a moisture sensor on the top of the hat, which the BBC micro:bit employs to trigger the warning. It also serves to fasten the BBC micro:bit to the hat, so look for thicker rather than thinner conductive thread to increase its strength.

- **Sewing materials**—A needle on its own is enough to build this project, but to make life easier and keep your fingers safe, consider adding a thimble, dedicated scissors, and a needle threader, as discussed earlier in this chapter.

- **A baseball cap**—The peak of a baseball cap makes for a great place to hang a BBC micro:bit, putting it in front of the wearer's eyes without blocking their vision too badly. It's probably a good idea to buy a cheap cap specifically for this project; you don't want to end up putting holes all over your favourite!

- **A ruler or measuring tape**—This allows you to make measurements, ensuring your BBC micro:bit and threads are positioned in the right places.

- **A fabric pencil**—You use this to mark out measurements on the baseball cap. You can also try using a normal pencil or pen, although these leave a permanent mark on your hat.

If you don't want to build the project right now, you can skip straight to the software section later in this chapter to experiment with the program. Touching the BBC micro:bit on Pin 0 while holding the GND pin is enough to simulate the rain sensor.

Building the Hat

Begin by measuring and writing down the distance between the centre of the hole of Pin 0 and the centre of the hole on the GND pin of your BBC micro:bit; it will be around 42mm (1.65"). This is the distance between the ending points of your two lengths of conductive thread, making it so that the thread can fasten the BBC micro:bit to the underside of the baseball cap's peak.

Find the centre point of the baseball cap's peak, and then make a small mark near the front of the peak on its underside. Measure half the distance you wrote down earlier (21mm/0.827") to the left and mark it; then do the same to the right (see Figure 12-8).

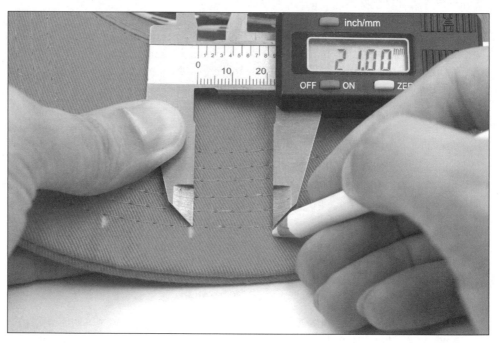

FIGURE 12-8: Marking the hat

Cut a length of conductive thread about four times longer than your original measurement—around 168mm (6.61"). Tie a knot in one end and thread the other end through the needle. Push the needle through the peak of the baseball cap from the underside where your left mark is. Pull it all the way through to the top, and then begin sewing a square wave pattern as shown in Figure 12-9 without passing the needle fully through the peak; instead, just push the needle sideways through the fabric above the peak to secure each part of the pattern. Stop when you reach the second mark, passing your needle through so the thread is closer to the front of the peak than the mark.

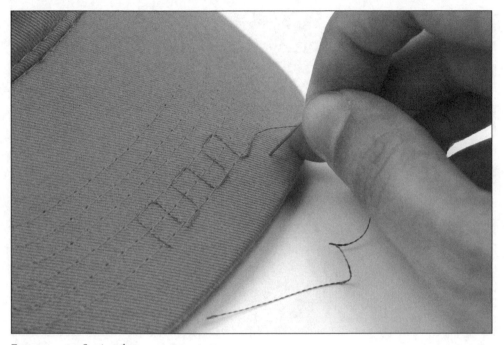

FIGURE 12-9: Sewing the square-wave pattern

Unthread your needle, leaving the rest of the thread hanging from the underside of the hat, and then cut a second length of conductive thread, knot it, and thread it through the needle. Place the needle on the right mark at the underside of the hat and begin the sewing process again, creating the same pattern but slightly above the old one (see Figure 12-10). Make sure that at no point do the two threads touch; they should be close, but not touching.

The spacing between the threads is important. Too far apart, and the rainwater isn't conductive enough for the sensor to work; too close together, and the threads connect to each other and short the circuit, causing the sensor to trigger even when no rain is present.

When you reach the end of the pattern, pass the needle fully through the peak and pull the thread through behind it, but don't unthread the needle just yet. Take your scissors and cut off any excess thread from the knotted ends of both conductive thread lengths, leaving them as close to flush with the peak of the baseball cap as possible.

FIGURE 12-10: Sewing the second square-wave pattern

Mounting the BBC micro:bit

The BBC micro:bit needs to be mounted to the hat, and you can use the lengths of thread hanging from the hat to do so while also making the electrical connections needed for the rain sensor. Place the BBC micro:bit upside down under the peak of the cap so that the LEDs are facing inward toward the wearer and the input-output pins are touching the peak. Holding it in place, position the BBC micro:bit so that the conductive thread you currently have threaded in the needle is touching the GND pin. Then begin sewing loops through the hole of the GND pin and the peak of the cap to hold it in place (see Figure 12-11) while taking care not to allow the thread to touch the other length of thread.

When you have finished, create a final loop and pass the thread through its own loop before pulling it tight to secure it, and then unthread the needle and cut off any excess thread. Rethread the needle using the remaining length of thread, and do the same to Pin 0 on the other side of the BBC micro:bit before again securing it in place and cutting off the excess.

FIGURE 12-11: Securing the BBC micro:bit

The Rain-Sensing Program

Start by opening your browser and going to makecode.microbit.org to load the JavaScript Blocks Editor, and then click the Editor toggle to switch to JavaScript Editor mode. Delete everything from the program listing, and then type in the following:

```
basic.forever(() => {
    basic.clearScreen()
    if (input.pinIsPressed(TouchPin.P0)) {
        basic.showLeds(`
            . . # # #
            . . # . #
            . . # . .
            # # # # #
            . # # # .
            `)
        basic.pause(500)
        basic.showIcon(IconNames.Chessboard)
        basic.pause(500)
    }
})
```

This small program is the entirety of the wearable's logic. Line 1 places the BBC micro:bit in an infinite loop, while Line 2 clears the screen of anything it is currently displaying. Line 3 checks to see if Pin 0 has been touched, and if it has it displays an image of an umbrella on the screen, pauses for 500 milliseconds, switches to a built-in icon of a chessboard which handily doubles as rain, and pauses for another 500 milliseconds before going back to the start of the loop, clearing the screen, and checking Pin 0 again.

It may seem strange to be testing Pin 0 for having been touched, but testing for rain and testing for a touch work in the same way: they both rely on conductivity and resistance. With one of your conductive threads connected to the GND pin and the other connected to Pin 0, rain completes the circuit and acts as a 'touch' just the same as your finger would.

There's a reason for the umbrella being upside down, too. With the BBC micro:bit connected to the peak of the cap by its input-output ports, it's upside down; any picture you display on the BBC micro:bit's display in this manner is also upside down as a result, which means you can't use the built-in 'umbrella' icon. Instead, the `basic.showLeds` instruction is used to create a custom image, which is simply the existing umbrella icon flipped upside down—making it seem the right way up to the person wearing the hat.

Click the Download button and flash the program onto your BBC micro:bit. To test if it works, you can either touch the conductive threads on the top of your cap with your finger or sprinkle a little water on them to simulate rain. If everything's working, the BBC micro:bit begins to alternate between the umbrella and rain images. If the images appear before you've pressed or wet the conductive threads, the threads are touching and creating a short; look closely to find out where, and then adjust them so they're no longer touching each other.

WARNING   Electricity and water don't mix well. Although it's perfectly safe to get your cap a little wet while testing the rain sensor, make sure that you don't accidentally soak the BBC micro:bit located underneath the peak. If the BBC micro:bit gets wet while it's powered on, it's likely to short out and break.

Battery Power

The only thing left to do to finish your wearable is to power it. Disconnect the micro-USB cable, if it is currently connected, and then take your battery pack and insert fresh batteries, making sure to put them in the right way around. Take the JST connector on the end of the battery pack's wires and insert it into the battery connector on the BBC micro:bit, making sure it's the right way around. The red, positive, wire goes to where the small + symbol is on the board, and the black, negative, wire goes to the – symbol. The connector is keyed to prevent you from putting it in backward; if it doesn't seem to be fitting properly, double-check that it's the right way up and try again.

The wires on the battery pack are long enough that you can, if you want, leave the battery pack loose and place it under the cap itself when you're wearing it. If you'd prefer something more permanent, you can attach it to the peak of the cap with nonconductive thread or use adhesive hook and loop fastener.

With the battery pack in place, you can wear your rain-sensing hat with pride (see Figure 12-12), knowing that you'll always be among the first to know if the clouds burst.

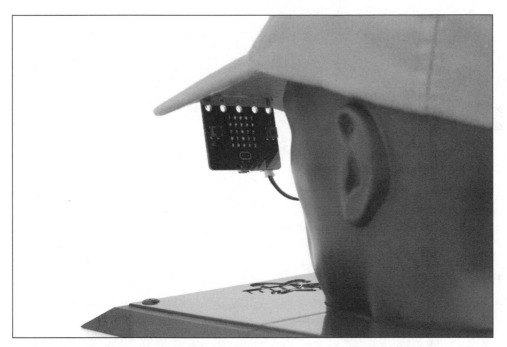

FIGURE 12-12: The finished rain-sensing hat

Chapter 13
Additional Resources

In this chapter

- A guide to the official resources available from the Micro:bit Educational Foundation

- Where to find lesson plans and guides for teachers and other educators

- An introduction to Code Clubs and how they can help students come to grips with the BBC micro:bit and coding

WHILE THIS BOOK is designed to get you up and running with the BBC micro:bit as quickly as possible, it can only take you so far. Eventually you're going to need to seek additional resources to continue your journey—and the good news is that there are plenty to be found, ranging from the official BBC micro:bit website to in-person meetups taking place around the globe.

> This chapter serves as a reference guide for some of the more popular micro:bit resources available, but it is by no means exhaustive. Searching for 'micro:bit' on your favourite search engine brings up plenty more resources. **NOTE**

The Micro:bit Educational Foundation

The Micro:bit Educational Foundation was set up in 2016 with one goal in mind: supporting the use of the BBC micro:bit in education throughout the world. Encompassing materials developed by and for the BBC as well as new resources developed since, the Foundation is a not-for-profit organisation entirely dedicated to the BBC micro:bit and a fantastic first stop for a wide range of resources.

If you've been working through the book from the start, the official website at `microbit.org` should be entirely familiar; it's where you can find the programming environments you've been using to work with your BBC micro:bit. As well as guides to using the BBC micro:bit at home and at school, the site has additional resources split into categories available by clicking the three-line 'hamburger' icon at the top-right: Let's Code, Ideas, Meet micro:bit, Teach, and a link to buy the BBC micro:bit from international resellers (see Figure 13-1).

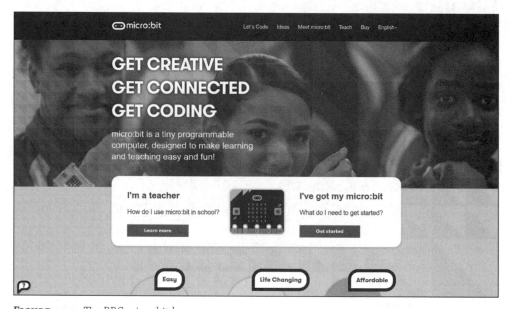

FIGURE 13-1: The BBC micro:bit homepage

Meet micro:bit offers a range of guides and animations for beginners, starting with a similar tour of the BBC micro:bit hardware that you can find in Chapter 1, 'Meet the BBC micro:bit', and working through writing programs in a variety of languages and connecting your BBC micro:bit to a battery pack for portable use.

Let's Code is where you can find the program editors discussed starting in Chapter 4, 'Programming Languages'. The editors aren't the only things available on this page, however. Below the Let's Code buttons which launch each editor is a Lessons button which takes you to language-specific guides and tutorials suitable from absolute beginners upward. Some languages also feature a Reference button which links to a more technical reference, handy for looking up how a particular feature or instruction works.

The Ideas page is where the BBC micro:bit website really gets interesting. Constantly updated, the Ideas section includes a round-up of projects, competitions, news stories, and educational materials from third-party websites including the BBC, Tech Will Save Us, Microsoft, and others. If you're struggling to come up with a concept for your next project, the Ideas page is sure to help.

The Teach category is more specific. As well as a rundown of the benefits offered by the BBC micro:bit to schools, colleges, and universities around the world, each link here provides teaching resources, hosted both by the Micro:bit Educational Foundation itself and specially-chosen third parties. More details on these resources can be found in the 'Official Teaching Resources' section later in this chapter.

The major advantage of the official BBC micro:bit website over finding third-party resources directly is a guarantee of quality. Everything listed on the website has been checked and vetted by the Foundation for technical accuracy and suitability, making it a great resource for younger students.

Official Teaching Resources

The BBC micro:bit has been designed to support curricula around the globe and ranging from the direct action of programming the device for coding lessons to its use in cross-curricular activities like art and music. To help teachers get the most from the BBC micro:bit, a wide range of teaching resources is available—starting with lesson plans and supporting materials developed by the Micro:bit Educational Foundation itself, which can be found on `microbit.org/teach` (see Figure 13-2).

FIGURE 13-2: Teaching resources on microbit.org

The Teach page offers links to internal and third-party resources designed specifically with the needs of teachers in mind. Some are linked to other resources designed to keep students entertained—such as lesson plans which tie in to BBC and Children's BBC programmes like the teen drama *Wolfblood* or the classic *Dr. Who*—while others stand alone.

Many projects listed on the site come with full lesson plans, including teachers' notes and a ready-written and compiled program which can be flashed onto any BBC micro:bit. You'll also find sample projects which link directly to a specific curriculum, such as the Key Stage 3 (KS3) Programme of Study for Computing used in the United Kingdom.

Third-Party Teaching Resources

While the Micro:bit Educational Foundation should be your first stop for teaching and training materials, it's far from the only place you can find useful information. The BBC micro:bit has been adopted by organisations around the world, and many have developed publicly-available resources to support the project.

TIP When browsing through third-party resources, you are likely to come across some which have been written with older versions of the BBC micro:bit's editors in mind. Developed by partners in the BBC micro:bit project and from which the current editors evolved, these older versions are called Microsoft Blocks, TouchDevelop, and CodeJungdoms JavaScripts. While these older editors are still available from the Let's Code section of the Micro:bit Educational Foundation website at the time of writing, they are being phased out in favour of the new, more powerful editors detailed in this book.

Programs written in Microsoft Blocks and Touch Develop can be loaded into the new editor and will be automatically converted to the latest version. Be aware, though, that screenshots and documentation provided with these programs will not match what you see in the new editor versions.

The Institution of Engineering and Technology

The Faraday educational arm of the Institution of Engineering and Technology (IET) has released a range of resources for the BBC micro:bit, including full lesson plans covering everything from a prototype energy-saving LED lighting system to a heart-rate monitor (see Figure 13-3). Designed primarily for the 11–14 age range, each lesson comes complete with an overview document, activity sheet, presentation, and student handout—and they're all provided entirely free of charge.

The IET has also provided more detailed case studies from industry members demonstrating how the BBC micro:bit can be used to prototype and simulate real-world electronic projects from rockets to Mars through to sports timing. These case studies come with a full booklet, sample program code, and a supporting video.

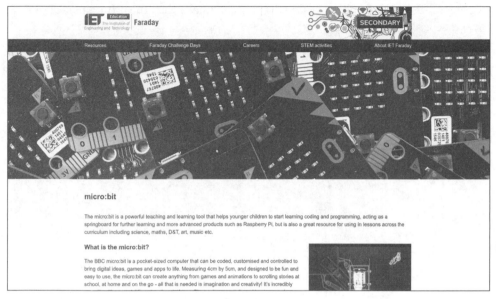

FIGURE 13-3: Teaching resources from the IET

You can find these resources at `faraday-secondary.theiet.org/stem-activities/microbit/`.

Computing At School

Part of BCS, the Chartered Institute for IT, Computing At School (CAS) has developed a range of materials to support the use of the BBC micro:bit in education (see Figure 13-4). Like the IET's lessons, the majority are aimed at the 11–14 age range, and all are made available completely free of charge to members and nonmembers alike.

The CAS lessons cover the most common languages for the BBC micro:bit, with a slight bias toward Python. Individual lessons and full schemes of work are available, and each is licensed under the permissive Creative Commons Attribution-Share Alike 3.0 licence, making them free for reproduction and reuse. The same site also offers materials for supporting devices, including the Raspberry Pi.

You can find these resources by going to `community.computingatschool.org.uk/resources/`, clicking on Language/Platform in the left Filters menu, and then clicking on BBC micro:bit.

Micro:bit for Primary Schools

Where many other sites focus on the 11–14 age range, Neil Rickus's Micro:bit for Primary Schools (MB4PS) offers resources supporting the use of the BBC micro:bit from 7 upwards (see Figure 13-5). As a result, you'll find the projects simpler than those from the IET or CAS, though no less engaging.

FIGURE 13-4: Teaching resources from CAS

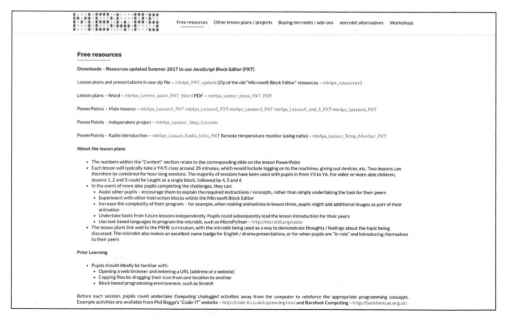

FIGURE 13-5: Teaching resources from MB4PS

The lesson plan booklet covers six introductory lessons, beginning with a basic introduction to the hardware as you can find in Chapter 1, and working through increasingly complex projects culminating in an independent project to build a step counter powered by the BBC micro:bit's accelerometer. Sample code, written in the child-friendly Block Editor, is provided at each step. All the materials are licensed as Creative Commons Attribution Non-Commercial Share Alike 4.0, allowing for their reproduction for teaching purposes but not for commercial reuse.

You can find these resources at `mb4ps.co.uk/resources`.

TES Magazine

Formerly know as the Times Educational Supplement, *TES Magazine* runs a website where contributors are able to upload resources, ranging from lesson plans through to colour posters and 'cheat sheets' for quick reference (see Figure 13-6).

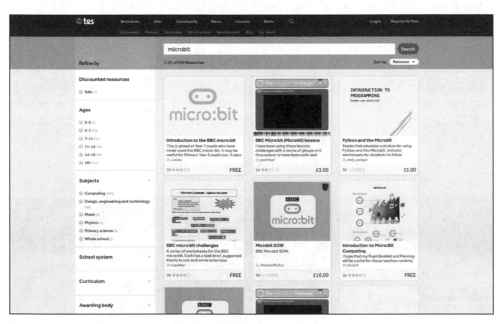

FIGURE 13-6: Teaching resources from *TES Magazine*

While membership to the TES website is free, not all the resources are. The magazine runs a resources marketplace of sorts, and contributors can choose a fee to associate with their materials. These fees are typically small—around £2–£5—and many contributors make their materials available free of charge, meaning it's still a worthwhile resource even for those operating their classrooms on a strictly restricted budget.

You can find these resources at `www.tes.com/resources`.

Code Clubs

A not-for-profit organisation founded in 2012 and brought under the umbrella of the Raspberry Pi Foundation in 2015, Code Club aims to support extracurricular education through a network of volunteers and educators who run coding clubs for children between 9 and 11 in more than 80 countries and 15 languages (see Figure 13-7). While the Code Club project itself supports a range of languages and hardware, many clubs have adopted the BBC micro:bit as the perfect device for introducing coding concepts.

If you're interested in attending a Code Club, you can find your nearest at `www.codeclub.org.uk/about`. If there aren't any near you, consider setting one up. As a volunteer-driven organisation, Code Clubs are entirely dependent on people finding a gap and filling it—and you don't need to be an expert yourself, thanks to a wealth of resources and the support of the Code Club community behind you.

If you'd rather look through the Code Club resources in the comfort of your own home, or use them in a more formal educational environment, you can find them at `www.codeclub projects.org`. All resources are free, though licensed for noncommercial use only.

As your skills improve, you can easily give back to the Code Club community should you so wish. As you work through the projects, either at a Code Club or on your own, you're given the opportunity to provide feedback or suggest amendments. When your skills are at their peak, you can even submit your own projects for inclusion—and if you are multilingual, you can help by translating existing projects and resources for international use.

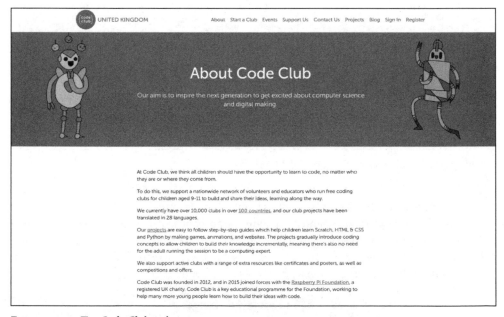

FIGURE 13-7: The Code Club website

Part IV

Appendices

Appendix A
JavaScript Blocks Recipes

THIS APPENDIX CONTAINS complete code listings for the programs detailed in Chapter 6, 'JavaScript', and Chapter 8, 'The Wireless BBC micro:bit'. These can be entered into the JavaScript Blocks Editor visually. Or, you can enter the alternative versions from Appendix B, 'JavaScript Recipes', in JavaScript Mode, and then use the Editor toggle to switch back to JavaScript Blocks Mode (see Chapter 6.)

Chapter 6: Hello, World! (Non-looping)

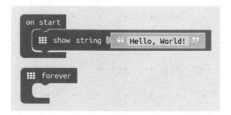

Chapter 6: Hello, World! (Looping)

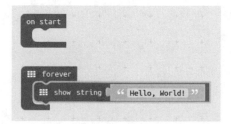

Chapter 6: Button Inputs (Single Button)

Chapter 6: Button Inputs (Two Buttons)

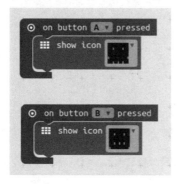

Chapter 6: Touch Inputs

Chapter 6: Temperature Sensor (No Formatting)

Chapter 6: Temperature Sensor (with Formatting)

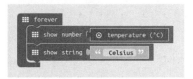

Chapter 6: Compass Sensor

Chapter 6: Accelerometer Sensor (Single Icon)

Chapter 6: Accelerometer Sensor (Two Icons)

Chapter 6: Accelerometer Sensor Data

Chapter 6: Fruit Catcher Game

Chapter 8: One-to-One Communication (BBC micro:bit A)

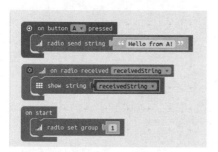

Chapter 8: One-to-One Communication (BBC micro:bit B)

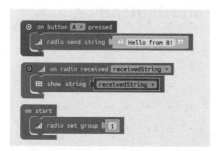

Chapter 8: One-to-Many Communication (BBC micro:bit C)

Chapter 8: Radio Groups Communication (BBC micro:bit A)

```
on button A pressed
    radio send string " Hello from A! "

on button B pressed
    radio set group 2
    show string " Switching to Group 2 "

on radio received receivedString
    show string receivedString

on start
    radio set group 1
```

Chapter 8: Radio Groups Communication (BBC micro:bit B)

```
on button A pressed
    radio send string " Hello from B! "

on button B pressed
    radio set group 2
    show string " Switching to Group 2 "

on radio received receivedString
    show string receivedString

on start
    radio set group 1
```

Chapter 8: Radio Groups Communication (BBC micro:bit C)

Chapter 10: Reading a Button Input

Chapter 10: Writing to an LED Output

```
forever
    digital write pin P1 ▾ to   1
    pause (ms)   1000
    digital write pin P1 ▾ to   0
    pause (ms)   1000
```

Chapter 10: Fading an LED via PWM

```
on start
    analog set period pin P1 ▾ to (µs)   10000

forever
    for index ▾ from 0 to   1023
    do
        analog write pin P1 ▾ to   index ▾
        pause (ms)   1
    for index ▾ from 0 to   1023
    do
        analog write pin P1 ▾ to   1023   - ▾   index ▾
        pause (ms)   1
```

Chapter 10: Reading an Analogue Input

```
forever
    show number   analog read pin P2 ▾
    show string   " "
```

Appendix B
JavaScript Recipes

THIS APPENDIX CONTAINS complete code listings for the programs detailed in Chapter 6, 'JavaScript', and Chapter 8, 'The Wireless BBC micro:bit', presented without comments to make them easier to type in and compare against. When a line of code would extend past the border of the page, a ↵ symbol is printed. When you see this symbol, continue to type the code without pressing the Enter or Return keys. If you're not sure how a line of code should be entered, visit the website at www.wiley.com/go/bbcmicrobituserguide to download plain-text versions of each program; these can then be used for reference or even simply copy and pasted directly into the editors.

Chapter 6: Hello, World! (Non-looping)

```
basic.forever(() => {

})
basic.showString("Hello, World!")
```

Chapter 6: Hello, World! (Looping)

```
basic.forever(() => {
    basic.showString("Hello, World!")
})
```

Chapter 6: Button Inputs (Single Button)

```
input.onButtonPressed(Button.A, () => {
    basic.showIcon(IconNames.Happy)
})
```

Chapter 6: Button Inputs (Two Buttons)

```
input.onButtonPressed(Button.A, () => {
    basic.showIcon(IconNames.Happy)
})
input.onButtonPressed(Button.B, () => {
    basic.showIcon(IconNames.Sad)
})
```

Chapter 6: Touch Inputs

```
let touches = 0
input.onPinPressed(TouchPin.P0, () => {
    touches += 1
    basic.showNumber(touches)
})
```

Chapter 6: Temperature Sensor (No Formatting)

```
basic.forever(() => {
    basic.showNumber(input.temperature())
})
```

Chapter 6: Temperature Sensor (with Formatting)

```
basic.forever(() => {
    basic.showNumber(input.temperature())
    basic.showString(" Celsius")
})
```

Chapter 6: Compass Sensor

```
basic.forever(() => {
    basic.showString("Heading " +↵
input.compassHeading())
})
```

Chapter 6: Accelerometer Sensor (Single Icon)

```
input.onGesture(Gesture.Shake, () => {
    basic.showIcon(IconNames.Surprised)
})
```

Chapter 6: Accelerometer Sensor (Two Icons)

```
input.onGesture(Gesture.Shake, () => {
    basic.showIcon(IconNames.Surprised)
    basic.pause(1000)
    basic.showIcon(IconNames.Asleep)
})
```

Chapter 6: Accelerometer Sensor Data

```
basic.forever(() => {
    basic.showString("X:" +↵
input.acceleration(Dimension.X))
    basic.showString("Y:" +↵
input.acceleration(Dimension.Y))
    basic.showString("Z:" +↵
input.acceleration(Dimension.Z))
})
```

Chapter 6: Fruit Catcher Game

```
let delay = 1000
let fruit: game.LedSprite = null
let player: game.LedSprite = game.createSprite(2, 4)
game.setScore(0)

basic.forever(() => {
    fruit = game.createSprite(Math.random(5), 0)
    basic.pause(delay)
    while (fruit.get(LedSpriteProperty.Y) < 4) {
        fruit.change(LedSpriteProperty.Y, 1)
        basic.pause(delay)
    }
```

```
    if (player.isTouching(fruit)) {
        game.addScore(1)
    } else {
        game.gameOver()
    }
    fruit.set(LedSpriteProperty.Brightness, 0)
    delay = delay - delay / 10
})
input.onButtonPressed(Button.A, () => {
    if (player.get(LedSpriteProperty.X) > 0) {
        player.change(LedSpriteProperty.X, -1)
    }
})

input.onButtonPressed(Button.B, () => {
    if (player.get(LedSpriteProperty.X) < 4) {
        player.change(LedSpriteProperty.X, 1)
    }
})
```

Chapter 8: One-to-One Communication (BBC micro:bit A)

```
radio.setGroup(1)
input.onButtonPressed(Button.A, () => {
    radio.sendString("Hello from A!")
})
radio.onDataPacketReceived(({receivedString}) => {
    basic.showString(receivedString)
})
```

Chapter 8: One-to-One Communication (BBC micro:bit B)

```
radio.setGroup(1)
input.onButtonPressed(Button.A, () => {
    radio.sendString("Hello from B!")
})
radio.onDataPacketReceived(({receivedString}) => {
    basic.showString(receivedString)
})
```

Chapter 8: One-to-Many Communication (BBC micro:bit C)

```
radio.setGroup(1)
input.onButtonPressed(Button.A, () => {
    radio.sendString("Hello from C!")
})
radio.onDataPacketReceived(({receivedString}) => {
    basic.showString(receivedString)
})
```

Chapter 8: Radio Groups Communication (BBC micro:bit A)

```
radio.setGroup(1)
input.onButtonPressed(Button.A, () => {
    radio.sendString("Hello from A!")
})
radio.onDataPacketReceived(({receivedString}) => {
    basic.showString(receivedString)
})
input.onButtonPressed(Button.B, () => {
    radio.setGroup(2)
    basic.showString("Switching to Group 2")
})
```

Chapter 8: Radio Groups Communication (BBC micro:bit B)

```
radio.setGroup(1)
input.onButtonPressed(Button.A, () => {
    radio.sendString("Hello from B!")
})
radio.onDataPacketReceived(({receivedString}) => {
    basic.showString(receivedString)
})
input.onButtonPressed(Button.B, () => {
    radio.setGroup(2)
    basic.showString("Switching to Group 2")
})
```

Chapter 8: Radio Groups Communication (BBC micro:bit C)

```
radio.setGroup(1)
input.onButtonPressed(Button.A, () => {
    radio.sendString("Hello from C!")
})
radio.onDataPacketReceived(({receivedString}) => {
    basic.showString(receivedString)
})
input.onButtonPressed(Button.B, () => {
    radio.setGroup(2)
    basic.showString("Switching to Group 2")
})
```

Chapter 10: Reading a Button Input

```
basic.forever(() => {
    while (pins.digitalReadPin(DigitalPin.P0) == 1) {
        basic.showIcon(IconNames.Surprised)
    }
    basic.showIcon(IconNames.Asleep)
})
```

Chapter 10: Writing to an LED Output

```
basic.forever(() => {
    pins.digitalWritePin(DigitalPin.P1, 1)
    basic.pause(1000)
    pins.digitalWritePin(DigitalPin.P1, 0)
    basic.pause(1000)
})
```

Chapter 10: Fading an LED via PWM

```
pins.analogSetPeriod(AnalogPin.P1, 10000)
let index = 0
basic.forever(() => {
    for (let index = 0; index <= 1023; index++) {
        pins.analogWritePin(AnalogPin.P1, index)
        basic.pause(1)
    }
    for (let index = 0; index <= 1023; index++) {
        pins.analogWritePin(AnalogPin.P1, 1023 - index)
        basic.pause(1)
    }
})
```

Chapter 10: Reading an Analogue Input

```
basic.forever(() => {
    basic.showNumber(pins.analogReadPin(AnalogPin.P2))
    basic.showString("")
})
```

Appendix C
Python Recipes

THIS APPENDIX CONTAINS complete code listings for the programs detailed in Chapter 7, 'Python', Chapter 8, 'The Wireless BBC micro:bit', Chapter 9, 'The BBC micro:bit and the Raspberry Pi', and Chapter 10, 'Building Circuits', presented without comments to make them easier to type in and compare against. When a line of code would extend past the border of the page, a ↵ symbol is printed. When you see this symbol, continue to type the code without pressing the Enter or Return keys. If you're not sure how a line of code should be entered, visit the website at www.wiley.com/go/bbcmicrobituserguide to download plain-text versions of each program; these can then be used for reference or even simply copy and pasted directly into the editors.

Chapter 7: Hello, World! (Non-looping)

```python
from microbit import *
display.scroll('Hello, World!')
```

Chapter 7: Hello, World! (Looping)

```python
from microbit import *
while True:
    display.scroll('Hello, World!')
```

Chapter 7: Button Inputs (Single Button)

```python
from microbit import *
while True:
    if button_a.is_pressed():
        display.show(Image.HAPPY)
```

Chapter 7: Button Inputs (Two Buttons)

```
from microbit import *
while True:
    if button_a.is_pressed():
        display.show(Image.HAPPY)
    if button_b.is_pressed():
        display.show(Image.SAD)
```

Chapter 7: Touch Inputs

```
from microbit import *
touches = 0
while True:
    if pin0.is_touched():
        touches += 1
        display.scroll(str(touches))
```

Chapter 7: Temperature Sensor (No Formatting)

```
from microbit import *
while True:
    display.scroll(str(temperature()))
```

Chapter 7: Temperature Sensor (with Formatting)

```
from microbit import *
while True:
    display.scroll(str(temperature()))
    display.scroll(' Celsius')
```

Chapter 7: Compass Sensor

```
from microbit import *
while True:
    display.scroll('Heading %s' % compass.heading())
```

Chapter 7: Accelerometer Sensor (Single Icon)

```python
from microbit import *
while True:
    if accelerometer.is_gesture("shake"):
        display.show(Image.SURPRISED)
```

Chapter 7: Accelerometer Sensor (Two Icons)

```python
from microbit import *
while True:
    if accelerometer.is_gesture("shake"):
        display.show(Image.SURPRISED)
        sleep(1000)
        display.show(Image.ASLEEP)
```

Chapter 7: Accelerometer Sensor Data

```python
from microbit import *
while True:
    display.scroll(str('X:%s' % accelerometer.get_x()))
    display.scroll(str('Y:%s' % accelerometer.get_y()))
    display.scroll(str('Z:%s' % accelerometer.get_z()))
```

Chapter 7: Fruit Catcher Game

```python
from microbit import *
import random
delay = 10
delayCounter = 0
playerPosition = [2, 4]
score = 0

while True:
    fruitPosition = [random.randrange(0,4), 0]
    while fruitPosition[1] <= 4:
        while delayCounter < delay:
            if button_a.was_pressed() and↵
```

```
(playerPosition[0] > 0):
                playerPosition[0] -= 1
            if button_b.was_pressed() and↵
(playerPosition[0] < 4):
                playerPosition[0] += 1
            display.clear()
            display.set_pixel(fruitPosition[0],↵
fruitPosition[1], 9)
            display.set_pixel(playerPosition[0],↵
playerPosition[1], 9)
            delayCounter += 1
            sleep(100)
        delayCounter = 0
        fruitPosition[1] += 1
    if fruitPosition[0] == playerPosition[0]:
        score += 1
        delay -= (delay / 10)
    else:
        display.scroll(('GAME OVER   SCORE %s' % score),
  loop=True)
```

Chapter 8: One-to-One Communication (BBC micro:bit A)

```
from microbit import *
import radio

radio.config(group=1)
radio.on()

while True:
    if button_a.was_pressed():
        radio.send('Hello from A!')
    message = radio.receive()
    if message != None:
        display.scroll(str(message))
```

Chapter 8: One-to-One Communication (BBC micro:bit B)

```
from microbit import *
import radio

radio.config(group=1)
radio.on()

while True:
    if button_a.was_pressed():
        radio.send('Hello from B!')
    message = radio.receive()
    if message != None:
        display.scroll(str(message))
```

Chapter 8: One-to-Many Communication (BBC micro:bit C)

```
from microbit import *
import radio

radio.config(group=1)
radio.on()

while True:
    if button_a.was_pressed():
        radio.send('Hello from C!')
    message = radio.receive()
    if message != None:
        display.scroll(str(message))
```

Chapter 8: Radio Groups Communication (BBC micro:bit A)

```
from microbit import *
import radio

radio.config(group=1)
radio.on()
```

```
while True:
    if button_a.was_pressed():
        radio.send('Hello from A!')
    if button_b.was_pressed():
        radio.config(group=2)
        display.scroll('Switching to Group 2')
    message = radio.receive()
    if message != None:
        display.scroll(str(message))
```

Chapter 8: Radio Groups Communication (BBC micro:bit B)

```
from microbit import *
import radio

radio.config(group=1)
radio.on()

while True:
    if button_a.was_pressed():
        radio.send('Hello from B!')
    if button_b.was_pressed():
        radio.config(group=2)
        display.scroll('Switching to Group 2')
    message = radio.receive()
    if message != None:
        display.scroll(str(message))
```

Chapter 8: Radio Groups Communication (BBC micro:bit C)

```
from microbit import *
import radio

radio.config(group=1)
radio.on()

while True:
    if button_a.was_pressed():
        radio.send('Hello from C!')
```

```
if button_b.was_pressed():
    radio.config(group=2)
    display.scroll('Switching to Group 2')
message = radio.receive()
if message != None:
    display.scroll(str(message))
```

Chapter 9: Reading the Accelerometer (BBC micro:bit)

```
from microbit import *
while True:
    x, y, z = accelerometer.get_x(),↵
accelerometer.get_y(), accelerometer.get_z()
    print(x, y, z)
    sleep(500)
```

Chapter 9: Reading the Accelerometer (Raspberry Pi)

```
import serial
ser = serial.Serial("/dev/ttyACM0", 115200, timeout=1)
ser.close()
ser.open()
while True:
    accelerometerData = ser.readline()
    print(accelerometerData)
```

Chapter 9: The BBC micro:bit as a Display

```
import serial, time
ser = serial.Serial("/dev/ttyACM0", 115200, timeout=1)
ser.close()
ser.open()
ser.write("from microbit import * \r".encode())
while True:
    ser.write("display.scroll('Hello, world!') \r".encode())
    time.sleep(10)
```

Chapter 9: A CPU Monitor

```
import serial, psutil, time
gradients = 20
readingList = [0,1,2,3,4]
ser = serial.Serial("/dev/ttyACM0", 115200, timeout=1)
ser.close()
ser.open()

print("Started monitoring system statistics for↵
micro:bit display.")
ser.write("from microbit import * \r".encode())
time.sleep(0.1)
ser.write("display.clear() \r".encode())
time.sleep(0.1)

barGraph = [[0, 0, 0, 0, 0], [0, 0, 0, 0, 0], [0,↵
   0, 0, 0, 0], [0, 0, 0, 0, 0], [0, 0, 0, 0, 0]]

while True:
    sysLoad = psutil.cpu_percent(interval=0)
    readingList.insert(0,int(sysLoad))
    del readingList[5:]
    for x in range(5):
        for y in range(5):
            readingComparison = (y+1) * gradients
            if (readingList[x] >= readingComparison):
                barGraph[y][x] = 9
            else:
                barGraph[y][x] = 0
ser.write("BARGRAPH = Image↵
(\"%s:%s:%s:%s:%s\") \r".encode() %↵
(''.join(str(e) |for e in barGraph[0]), '↵
'.join(str(e) for e in barGraph[1]), '↵
'.join(str(e) for e in barGraph[2]), '↵
'.join(str(e) for e in barGraph[3]), '↵
'.join(str(e) for e in barGraph[4])))
    time.sleep(0.1)
    ser.write("display.show(BARGRAPH) \r".encode())
    time.sleep(0.9)
```

Chapter 10: Reading a Button Input

```python
from microbit import *
while True:
    while (pin0.read_digital() == 0):
        display.show(Image.SURPRISED)
    display.show(Image.ASLEEP)
```

Chapter 10: Writing to an LED Output

```python
from microbit import *
while True:
    pin1.write_digital(1)
    sleep(1000)
    pin1.write_digital(0)
    sleep(1000)
```

Chapter 10: Fading an LED via PWM

```python
from microbit import *
pin1.set_analog_period(1)
while True:
    for brightnessValue in range(0,1024,1):
        pin1.write_analog(brightnessValue)
        sleep(1)
    for brightnessValue in range(1023,-1,-1):
        pin1.write_analog(brightnessValue)
        sleep(1)
```

Chapter 10: Reading an Analogue Input

```python
from microbit import *
while True:
    potValue = pin2.read_analog()
    display.scroll(str(potValue))
```

Appendix D
Pin-Out Listing

THE FOLLOWING FIGURE is a numbered diagram of the BBC micro:bit's general-purpose input-output (GPIO) pins, both large and small. The table describes the functions available on each pin.

For more information on the BBC micro:bit GPIO pins, see Chapter 10, 'Building Circuits'.

Pin No.	Main Function	Additional Function	Notes
0	GPIO Pin 0	Analogue Input	Large pin, Weak Pull-Up Resistor
1	GPIO Pin 1	Analogue Input	Large pin, Weak Pull-Up Resistor
2	GPIO Pin 2	Analogue Input	Large Pin, Weak Pull-Up Resistor
3V	3V Power Supply		Large Pin
GND	Ground Connection		Large Pin
3	LED Column 1	GPIO Pin 3, Analogue Input	Small Pin
4	LED Column 2	GPIO Pin 4, Analogue Input	Small Pin
5	Button A Input	GPIO Pin 5	Small Pin, Pull-Up Resistor

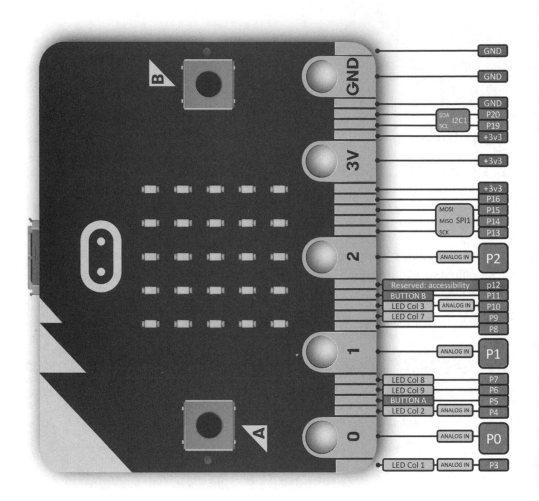

Pin No.	Main Function	Additional Function	Notes
6	LED Column 9	GPIO Pin 6	Small Pin
7	LED Column 8	GPIO Pin 7	Small Pin
8	GPIO Pin 8		Small Pin
9	LED Column 7	GPIO Pin 9	Small Pin
10	LED Column 3	GPIO Pin 10, Analogue Input	Small Pin
11	Button B Input	GPIO Pin 11	Small Pin, Pull-Up Resistor
12	Reserved for Accessibility		Small Pin
13	GPIO Pin 13	SPI1 SCK	Small Pin
14	GPIO Pin 14	SPI1 MISO	Small Pin
15	GPIO Pin 15	SPI1 MOSI	Small Pin
16	GPIO Pin 16		Small Pin
17	3V Power Supply		Small Pin, Linked to 3V
18	3V Power Supply		Small Pin, Linked to 3V
19	I^2C1 SCL	GPIO Pin 19	Small Pin, Linked to Accelerometer and Compass
20	I^2C1 SDA	GPIO Pin 20	Small Pin, Linked to Accelerometer and Compass
21	Ground Connection		Small Pin, Linked to GND
22	Ground Connection		Small Pin, Linked to GND

Pins 3, 4, 6, 7, 9, and 10 are all used to control the BBC micro:bit's LED matrix display; they may be used as GPIO pins (and in the cases of Pins 3, 4, and 10 as analogue inputs) when the display is not in use.

Pins 5 and 11 are linked to Button A and Button B, respectively. They are fitted with a pull-up resistor, meaning they are at 3V by default and must be brought low to register as an input.

The small pins should not be used except with a compatible breakout board to allow safe electrical connections to be made. Examples of these breakout boards can be found in Chapter 11, 'Extending the BBC micro:bit'.

When working with the BBC micro:bit's pins, be aware of the following limitations:

- You cannot draw more than 90mA from the 3V power supply pins.
- All GPIO pins are limited to 0.5mA by default, with a maximum of three pins configurable to a 5mA high-current mode at any given time.
- Pulse Width Modulation (PWM) output is available on a maximum of three pins at any given time.

More detail on the BBC micro:bit's pins and their functions can be found on the official website at `tech.microbit.org/hardware/edgeconnector_ds/`.

Index